# An Asset-Based Approach to Advancing Latina Students in STEM

This timely volume challenges the ongoing underrepresentation of Latina women in science, technology, engineering, and mathematics (STEM), and highlights resilience as a critical communal response to increasing their representation in degree programs and academic posts.

*An Asset-Based Approach to Advancing Latina Students in STEM* documents the racialized and gendered experiences of Latinas studying and researching in STEM in US colleges, and centers resilience as a critical mechanism in combating deficit narratives. Adopting an asset-based approach, chapters illustrate how Latinas draw on their cultural background as a source of individual and communal strength and indicate how this cultural wealth must be nurtured and used to inform leadership and policy to motivate, encourage, and support Latinas on the pathway to graduate degrees and successful STEM careers. By highlighting strategies to increase personal resilience and institutional retention of Latina women, the text offers key insights to bolstering diversity in STEM.

This text will primarily appeal to academics, scholars, educators, and researchers in the fields of STEM education. It will also benefit those working in broader areas of higher education and multicultural education, as well as those interested in the advancement of minorities inside and outside of academia.

**Elsa M. Gonzalez** is Assistant Professor of Higher Education at the University of Houston, USA.

**Frank Fernandez** is Assistant Professor of Higher Education at the University of Mississippi, USA.

**Miranda Wilson** earned a Ph.D. in Higher Education Leadership and Policy Studies at the University of Houston, USA.

# Routledge Research in STEM Education

The *Routledge Research in STEM Education* series is home to cutting-edge, upper-level scholarly studies and edited collections covering STEM Education.

Considering science, technology, engineering, and mathematics, texts address a broad range of topics including pedagogy, curriculum, policy, teacher education, and the promotion of diversity within STEM programmes.

Titles offer dynamic interventions into established subjects and innovative studies on emerging topics.

Books in this series include:

**An Asset-Based Approach to Advancing Latina Students in STEM**
Increasing Resilience, Participation, and Success
*Edited by Elsa M. Gonzalez, Frank Fernandez, and Miranda Wilson*

For more information about this series, please visit: https://www.routledge.com/Routledge-Research-in-STEM-Education/book-series/RRSTEM

# An Asset-Based Approach to Advancing Latina Students in STEM

Increasing Resilience, Participation, and Success

Edited by Elsa M. Gonzalez, Frank Fernandez, and Miranda Wilson

Routledge
Taylor & Francis Group

NEW YORK AND LONDON

First published 2021
by Routledge
605 Third Avenue, New York, NY 10017

and by Routledge
2 Park Square, Milton Park, Abingdon, Oxon, OX14 4RN

First issued in paperback 2022

*Routledge is an imprint of the Taylor & Francis Group, an informa business*

© 2021 Taylor & Francis

The right of Elsa M. Gonzalez, Frank Fernandez, and Miranda Wilson
to be identified as the authors of the editorial material, and of the
authors for their individual chapters, has been asserted in accordance
with sections 77 and 78 of the Copyright, Designs and Patents
Act 1988.

Publisher's Note
The publisher has gone to great lengths to ensure the quality of this
reprint but points out that some imperfections in the original copies may
be apparent.

*Library of Congress Cataloging-in-Publication Data*
Names: Gonzalez, Elsa M., editor.
Title: An asset-based approach to advancing Latina students in STEM:
   increasing resilience, participation, and success / edited by
   Elsa M. Gonzalez, Frank Fernandez, and Miranda Wilson.
Description: New York, NY : Routledge, 2021. | Series: Routledge
   research in STEM education | Includes bibliographical references
   and index.
Identifiers: LCCN 2020027917 | ISBN 9780367433758 (hardback)
   ISBN 9781003002758 (ebook)
Subjects: LCSH: Women in science—United States. | Hispanic
   American women—Education (Higher) | Science—Study and
   teaching (Higher)—United States. | Technology—Study and
   teaching (Higher)—United States. | Engineering—Study
   and teaching (Higher)—United States. | Mathematics—Study
   and teaching (Higher)—United States.
Classification: LCC Q130.A87 2021 | DDC 507.1/173—dc23
LC record available at https://lccn.loc.gov/2020027917

ISBN: 978-0-367-63098-0 (pbk)
ISBN: 978-0-367-43375-8 (hbk)
ISBN: 978-1-003-00275-8 (ebk)

DOI: 10.4324/9781003002758

Typeset in Bembo
by Apex CoVantage, LLC

To the Latinas in my family . . . my mom, my daughter, my mother in law, my sisters in law . . . to my Latino partners . . . my husband, my son, my dad, my brothers . . . to them and those who have showed me that resilience is part of who we are, thank you for being an essential part of the journey.

Elsa M. Gonzalez

For Elisa Raquel Fernandez, who reminds me every day that first-generation Latinas succeed in multiple STEM fields.

Frank Fernandez

For Latinas who dared to dream of conquering STEM . . . may they continue to inspire others and keep the dream alive.

Miranda Wilson

# Contents

# Acknowledgements

We would like to acknowledge several people; thanks to them, the edited volume *An Asset-Based Approach to Advancing Latina Students in STEM: Increasing Resilience, Participation, and Success* is a reality. The idea started early in 2019, and despite several challenges, the volume is a finished product thanks to the support of friends and colleagues.

First, the edited volume would not be possible without the support, guidance, and thoughtful advice of our editorial team at Routledge: Ellie Wright and AnnaMary Goodall. Since our first conversation, Ellie has been friendly, helpful, and an excellent advocate for our work. When we first met Ellie, she had recently started her current role at Routledge—but you would have never known it. Ellie guided us through all the logistical matters so we could keep our focus on the content of the volume.

We also thank the several contributors of this edited volume for getting everything across the finish line during the COVID-19 global pandemic. Our contributors respected deadlines while balancing their many responsibilities and were strongly committed to this project. Several authors contributed to a research session at the 2020 American Association of Hispanics in Higher Education (AAHHE) conference. Their participation in that voluntary convening created synergy as authors shared feedback on chapter drafts and made final edits to their contributions to the volume. That group helped generate a sense of community and excitement—right before the nation embraced social distancing—which helped push this project to completion.

Elsa would like to thank Juan Carlos, Angelina, and Carlos Baltazar for your support, your confidence, and your love that keeps us strong and true to our values.

Finally, the following individuals provided thoughtful insights at different stages of the project: Yvonna S. Lincoln, Deborah Santiago, Paula Myrick Short, and Fred Bonner. Their advice, support, and ideas made this volume a stronger contribution and more likely to be helpful to administrators and others who can work toward improving equity in higher education.

Elsa M. Gonzalez
Frank Fernandez
Miranda Wilson

# Figures

# Tables

# Contributors

**Hilda Cecilia Contreras Aguirre** is Adjunct Professor for the First-Year Learning Communities program at Texas A&M University, Corpus Christi (TAMU-CC) and a Visiting Scholar with the Department of Educational Leadership & Policy Studies at the University of Houston. From 2012 to 2013, she served as an English Instructor for the Program of Extended Learning at Santa Barbara City College in Santa Barbara, California. From 2016 to 2019, she worked as a mentor for the First-Islander Scholar Academy program in the Center for Academic Student Achievement at TAMU-CC. Her research interests include Latinas' performance in STEM, mentoring and advising in college, and college minority students' persistence.

**Rosa Banda** is Assistant Professor of Educational Leadership at Texas A&M University, Corpus Christi. Dr. Banda's primary research interests include high-achieving Latinas in engineering, gifted poor Students of Color, faculty diversity, and qualitative research.

**Selyna Pérez Beverly** is a doctoral candidate and research assistant in the Center for the Study of Higher and Postsecondary Education at the University of Michigan. Her research interests include race and gender inequity in STEM, effects of instructors and peers on underrepresented populations in STEM, masculinity in engineering culture, and feminist research methodologies.

**Tamara T. Coronella** is an administrator at the Ira A. Fulton Schools of Engineering at Arizona State University. In her professional work, she leads student success, academic advising, and student engagement activities for graduate and undergraduate students studying both in online and immersion courses. Her research focuses on transformative leadership, sense of belonging, validation, and social and cultural capital with Latinas in engineering fields.

**Veronica Crawford** is a higher education doctoral student at the University of Mississippi. Ms. Crawford worked for seven years as an administrator for the Oral-Maxillofacial Surgery residency program at the University of Mississippi Medical Center School of Dentistry. Ms. Crawford also served as an e-learning administrator for dental student education. Her research interests focus on technological advancements in teaching practices and their impact on student learning outcomes.

**Araceli Espinoza–Wade** is a researcher and practitioner. She is a first-generation college student. Her work focuses on the college experiences and outcomes of racial and ethnic minority (REM) students, first-generation college students, and REM students in science, technology, engineering, and mathematics (STEM) fields.

**Frank Fernandez** is Assistant Professor of Higher Education and a fellow with the Sarah Isom Center for Women and Gender Studies at the University of Mississippi. He studies educational policy and equity issues. He is co-editor of *Affirmative Action and Racial Equity: Considering the Fisher Case to Forge the Path Ahead* (Routledge, 2015) and *The Century of Science: The Global Triumph of the Research University* (Emerald, 2017 [hardback]; 2019 [paperback]). He is a co-author of *The Contested Campus: Aligning Professional Values, Social Justice, and Free Speech* (NASPA, 2020).

**Ariana L. Garcia** is a doctoral student in the Department of Educational Psychology and Higher Education at the University of Nevada, Las Vegas. Ariana's research broadly explores access and equity issues facing higher education. While much of her work focuses on the educational and personal experiences of graduate Students of Color, she has also worked as an advisor, teacher, counselor, and in strategic planning in higher education.

**Elsa M. Gonzalez** is Assistant Professor of Higher Education in the Department of Educational Leadership and Policy Studies at the University of Houston. Previously, she held faculty positions at Texas A&M University, Corpus Christi and Texas A&M University, College Station. Dr. Gonzalez is Regional Editor of the *International Journal of Qualitative Studies in Education* and an editorial board member of the *Journal of Hispanic Higher Education* and the *Journal of Minority Achievement, Creativity, and Leadership*. She is the author of 101 publications. Her research interests include issues in higher education such as underrepresented students, Latina/o students, school–college and readiness, access, resilience, retention, and graduation in STEM fields; HSI institutions; higher education leadership; and methodological issues in cross-language and cross-cultural qualitative data analysis. She has professional and academic leadership experience as bilingual scholar in higher education institutions in Mexico, Spain, the United States, and China.

**Lidia Herrera-Rocha** is an assistant professor of instruction of bilingual education and a site coordinator at the University of Texas, El Paso. Her research focuses on persistence and resilience among Latinx students in STEM, bilingual education programs on the US–Mexico border, and emergent bilinguals' languaging practices and identities. As a site coordinator in the department of teacher education, she coaches and supports pre-service teachers during their yearlong residency at a local school campus. Her experiences as a former teacher and student in a bilingual education program, her studies in literacy and biliteracy, and her pilot research as a graduate student have led her to focus on emphasizing the experiences, challenges, and perspectives of Latinx youth through their own voices.

**Juanita K. Hinojosa** is pursuing her Ph.D. in Higher Education at the University of Nevada, Las Vegas (UNLV). Juanita identifies as a practitioner-scholar and currently serves as Assistant Director for Community Partnerships and Scholar Development in the Office of Service Learning and Leadership at UNLV. Juanita identifies as a first-generation Mexican immigrant who first came to this country as an undocumented child. She is passionate about researching, developing, and implementing inclusive practices that support transformative equity initiatives in higher education.

**Aurora Kamimura** is a fellow in the Office of the Vice Provost at Washington University of St. Louis and an organizational development consultant in the areas of higher education, diversity, equity, and inclusion. She was previously Visiting Assistant Professor at the University of North Texas. Her research agenda focuses on broadening access and equity in the P-20/professoriate pathway. More recently, her research has looked to identify promising practices for enhancing structural diversity in the STEM professoriate by focusing distinctly on the recruitment and hiring process. As an educational professional, Dr. Kamimura has over 15 years of experience in statewide college access initiatives and diversity, equity, and inclusion efforts. She recently served as an Associate Dean of Student Services, working with students in the K-20 pipeline.

**Sanga Kim** is a postdoctoral research fellow at The University of Texas at El Paso, the lead institution for the Computing Alliance for Hispanic-Serving Institutions (CAHSI) initiative. In 2018, CAHSI was named as one of only five of the National Science Foundation's inaugural INCLUDES (Inclusion across the Nation of Communities of Learners of Underrepresented Discovers in Engineering and Science) Alliances. Her research uses sociological perspectives to study diversity and equity issues in higher education, including college access, student success, and diversity in STEM education. Her current work focuses on the promotion of educational equity and social justice for Latinx students in computing fields. She has published and co-authored several articles published in the following journals: *Educational Evaluation and Policy Analysis, Educational Policy, Journal of Student Affairs Research and Practice,* and *Teachers College Record.*

**Charles Lu** is Director of the Office of Academic Support and Instructional Services at the University of California, San Diego. In his role as director, he provides strategic oversight and management of the center and works collaboratively to create programs and initiatives that provide a holistic and student-centered academic experience. Previously he was Director of the Gateway Scholars Program, the largest student success program at The University of Texas at Austin. Dr. Lu has served as an educational researcher, consultant, school director, academic coach, and middle school science teacher, and was the recipient of the Toyota International Teacher of the Year Award.

**Erika Mein** is Associate Dean of Undergraduate Studies and Educator Preparation in the College of Education at the University of Texas, El Paso and Associate Professor in the Department of Teacher Education. Her scholarship focuses on disciplinary literacies in postsecondary contexts, with a particular emphasis on the literacies and professional identity development of emergent bilingual and Latinx students. Her research has been published in journals such as *Theory into Practice*, *Action in Teacher Education*, and the *Journal of Hispanic Higher Education*.

**Mauricio Molina** earned a Ph.D. in Higher Education Leadership and Policy Studies at the University of Houston. Previously, he worked in academic affairs and advising at the University of Houston and Louisiana State University. His research interests include campus ecology, campus planning and design, higher education funding and capital outlay, Latinx student success, and HSIs.

**Helena Muciño Guerra** has interdisciplinary academic and professional experience in the field of education and music. She is a doctoral student in the Teaching, Learning, and Culture program in the College of Education at the University of Texas, El Paso. Her research focuses on teacher preparation of Latinx university students and identity development of Mexican and Mexican-origin students in engineering and computer science education. Helena teaches Spanish to medical students at Texas Tech University, which has brought an interest in researching the intersection of language practices and race.

**Daisy Ramirez** is a Ph.D. student in the Higher Education and Organizational Change program at the University of California, Los Angeles. Daisy's research interests include graduate school access and community college student trajectories, as well as equity and access in computer science fields for Students of Color, specifically Latina students. Daisy has been involved in various mixed-methods research projects and continues to present her research in various outlets.

**Blanca Rincón** is Assistant Professor of Higher Education at the University of Nevada, Las Vegas. Her research agenda is concerned with equity issues in higher education, with a specific focus on access and success for Latinx and other underserved students (e.g., women, low-income, first-generation, and Students of Color) in STEM (science, technology, engineering, and mathematics). Her research has been published in the *Journal of Hispanic Higher Education, Journal of International Qualitative Studies in Education, Journal of College Student Development*, and *Teachers College Record*.

**Hyun Kyoung Ro** is Associate Professor in the Higher Education Program at the University of North Texas. Her research interests include minoritized college student learning and outcomes; diversity issues in STEM education; program assessment in higher education and student affairs; and quantitative research methodologies. She has been the leading author on articles published in higher education and engineering education journals, such as *The Review of Higher Education, Research in Higher Education, Educational Policy, Journal of Engineering Education*, and *Journal of Women and Minorities in Science and Engineering*.

**Sarah L. Rodriguez** is Associate Professor of Higher Education & Learning Technologies at Texas A&M University, Commerce. Her research addresses equity, access, and retention in higher education, with a focus on Latina/o/x students in STEM. She has been involved with several large-scale interdisciplinary research projects funded by the National Science Foundation, the Kapor Center, and the Center for the Study of Community Colleges. She currently serves as co-principal investigator for a mixed-methods study on Latina students in undergraduate computing. She has authored multiple articles on STEM identity experiences and sense of belonging of Latina undergraduate students in STEM. Dr. Rodriguez was a 2018 Faculty Fellow with the American Association of Hispanics in Higher Education (AAHHE) and the recipient of the 2020 Barbara K. Townsend Emerging Scholar Award given by the Council for the Study of Community Colleges. To learn more about her current projects, visit http://sarahlrodriguez.com/.

**Sarah Churchill Turner** is a doctoral student in the Measurement, Quantitative Methods, and Learning Sciences program at the University of Houston. She recently presented posters at the Society of Personality and Social Psychology and Annual Houston Symposium for Research in Education and Psychology about optimism, hope, and well-being, with Dr. Sascha Hein in the Positive Outlook Study. The team compared the implicit measure of optimism and the explicit measures of these constructs, and then among Hispanic and non-Hispanic students. Her research interests include resilience, mental health, Latina/o first-generation college students, measurement techniques in health, and mixed methodology research in psychology and education.

**Kristan M. Venegas** is the LaFetra Endowed Professor of Excellence in Teaching and Learning in the LaFetra College of Education at the University of La Verne in La Verne, California. Her research focuses on financial aid and college access for low-income students and Students of Color. She also serves as a research associate in the University of Southern California's Pullias Center for Higher Education.

**Dina Verdín** is Assistant Professor of Engineering in the Polytechnic School of the Ira A. Fulton Schools of Engineering at Arizona State University. Her research strands focus on access, college transition, persistence, and identity congruence for underrepresented students in engineering (i.e., Latinas, first-generation college students). Her goal is to create ways for underrepresented students to see themselves as engineers and develop asset-based approaches to support their pathways into and through engineering. She has won several research awards, including the IEEE Frontiers in Education Conference 2018 Best Diversity Paper and the Purdue University College of Engineering Outstanding Graduate Student Research Award, and her dissertation was selected among the top three at the 2018 American Educational Research Association Annual Meeting at the Division D In-Progress Research Gala.

**Miranda Wilson** earned a Ph.D. from the Higher Education Leadership and Policy Studies program at the University of Houston. She successfully defended her dissertation, titled "The Impact of Texas Emerging Research Universities on Student Outcomes." Her research focuses on the ways that research universities and their prestige-seeking behaviors impact student access and success outcomes. She is the managing co-editor of the *Journal of Research on Leadership Education* and previously worked as a graduate assistant for a mentor group for Black women doctoral students.

# Foreword

Latinas are resilient and successful. And yet, their success in areas where they are not well represented belies the numbers of those we know could, but do not, succeed. One such area where Latinas' success is not generally discussed is science, technology, engineering, and math (STEM). STEM has been prioritized as a field of study since before the 1960s and the "space race." However, much less priority has been given to understanding who has earned STEM degrees, who has not, the characteristics and components of success, and how to tactically increase success in STEM for underrepresented groups, including Latinas. This book provides some of this much-needed attention and perspective on Latinas in STEM and offers an asset-based approach to inform and compel action.

I have been working to improve Latino student success in postsecondary education for over 20 years, and I co-founded *Excelencia* in Education (*Excelencia*) in 2004 to address the void in national and public policy discussions about the opportunities for advancing Latinos in higher education and closing gaps in educational attainment. When we started *Excelencia*, we noted that those involved in policy discussions and funding had a very deficit-based perspective of Latinos in education (if they had a perspective at all). The focus was on how poorly we were performing in schools based on the conventional measures and understanding of the time. As a group, Latinos did not look like traditional students, did not follow the traditional pathways, and did not make choices in traditional ways to go to traditional institutions. And this profile is even more prevalent today.

However, the majority of students today do not represent the traditional profile. They do not go straight from high school to college academically prepared, enroll full-time, live on campus, have parents with a college degree, work little to no hours, and complete in two or four years. This traditional profile represents less than 20% of college students today. The majority of students today are likely to need some additional college preparation, have mixed enrollment (full- and part-time), live at home or off-campus, be a first-generation college attendee, work 30 hours or more per week, and take more time to complete. Latino students represent a post-traditional profile of students. We use the term "post-traditional" because the more conventional term, "non-traditional" student, is a more deficit-based term insinuating that we are not the norm, rather than representing a profile of students that are the growing majority redefining the conventional under-standing of higher education and student success.

Over the last 16 years, *Excelencia*'s movement of national and policy con-versations from a deficit-based profile of Latino students to a more asset-based profile that can inform an asset-based approach to increasing Latino student success in higher education has also been key. When we started, there was lit-tle discussion of the strengths, resilience, perseverance, and value that Latinos have in education to help our families and our communities. This occurs more now. Further, as post-traditional students become the focus, moving from a priority focus on traditional students, pathways, and institutions to a post-traditional profile with Latinos as the trendsetters has also increased.

Understanding the current profile of Latinos in higher education overall, Latinas, and in STEM specifically, is key to framing an approach to increas-ing their resilience and success. Consider the following summary of Latinas in higher education and STEM.

Hispanics earned 13% of all bachelor's degrees awarded in the United States in 2017–2018. While this is lower than the overall representation of Hispanics in higher education (19%[1]), it still represents a significant increase in degree completion overall, and in representation of all degrees awarded. Twelve years prior, Hispanics had earned only 7% of all bachelor's degrees awarded (2004–2005); our representation has almost doubled.

Latinas have made great strides in postsecondary enrollment, degree com-pletion, and faculty representation over the last ten years that can be built upon in policy and practice. Hispanic females have increased their college enrollment by over 150% since 2000 and now represent 11% of all students and 59% of Hispanic students.[2] They also earned about 13% of all degrees awarded in 2017–2018. This represents 20% of degrees awarded to women and 68% of degrees awarded to Hispanic students that year.[3] Further, His-panic females represented 2% of all faculty, 5% of all female faculty, and 50% of Hispanic faculty in 2018.[4]

Looking more closely at STEM, Latinas have also made significant improvement in representation over the last ten years. More than ten years ago, I was fortunate enough to be asked to provide a data profile of Latinas in STEM for a publication, *Flor y ciencia: Chicanas in Science, Mathematics and Engineering*, edited by Norma E. Cantú. Back then, Latinas earned about 60% of all bachelor's degrees awarded to Hispanics but only 37% of degrees awarded to Hispanics in STEM fields (2004–2005). As I look to the most recent data, the representation has not changed. Latinas still earned 61% of bachelor's degrees awarded to Hispanics but earned only 38% of degrees awarded to Hispanics. However, the numbers of Latinas who have earned degrees in STEM have increased.

In 2004–2005, Latinas earned fewer than 5,000 bachelor's degrees in STEM. In 2017–2018, Latinas earned almost 17,000 bachelor's degrees in

STEM (10% of all bachelor's degrees awarded in STEM). Further, Hispanic females went from representing 8% of all degrees earned in STEM to 14% (2008–2009 to 2017–2018).[5] And growth in degrees earned within academic levels in STEM—from certificate to doctorate—has been similar. For example, at the associate level, Latinas increased their representation from 12% to 23% of degrees earned in STEM in those ten years. At the doctorate level, Latina increased their representation from 5% to almost 8% of degrees earned in those ten years. Further, the largest representation of Latinas is in the sciences (and more specifically, biological and biomedical sciences) and the lowest representation is in math.

We know that many Latinas have worked hard for this progress to occur, and many departments, institutions, and employers across the country have provided support to reach this level of progress. As laudable as this progress has been, it is worth noting that the baseline from which this progress has been made was relatively low. Latinas are still underrepresented in STEM compared to the Hispanic population overall, other women, and the growing needs for STEM professionals in higher education, industry, and the economy. Further, Latinas are significantly more likely to earn degrees in business, humanities, education, and health than in STEM.[6] It is clear that there is still more progress needed for Latinas in STEM to reach the levels of representation and attainment in faculty and all levels of education.

While this data profile of Latinas in STEM is useful for framing an approach to increasing their resilience and success, these data are only part of the picture to understanding effective strategies, practices, and policies for serving Latinas in STEM. An asset-based approach to increasing resilience and success is needed. How do we frame approaches to the recruitment, retention, and degree completion of Latinas in STEM? Persistence through preparation, resilience, and perseverance are concrete actions by Latinas to succeed in STEM addressed by several authors in this book.

## Asset-Based Approach and Focus

As noted in *Excelencia's* work and in the title of this book, an asset-based approach is needed when focused on Latinas and STEM. For too long, and by too many, there has been a deficit perspective on Latinos in higher education overall, and in Latinas specifically. This deficit perspective continues a focus on the negative, rather than the potential and opportunity to improve and change—and is amplified in STEM, where Latina representation is low compared to our representation in higher education overall.

The ignorance (not knowing) about the Latino community's strengths, value, and resilience continues in some conversations, but the voices being raised to expose and correct this ignorance has grown over the last 16 years. And the chapters in this book provide additional perspectives, experiences, voices, and opportunities to ensure a more accurate and asset-based approach to serving our Latino students in higher education and beyond. Authors also provide powerful *testimonios* (personal stories) of Latinas in STEM. These stories capture the perspectives of Latinas in STEM. They share the strengths and needs as well as the opportunities and challenges for making a significant difference in the lives of individuals, in the field, and in communities overall. However, the success of Latinas in STEM should not be solely dependent on their efforts. How must departments, institutions, community, and society help meet them part of the way? I would argue that to do so requires a tactical plan and intentionality in serving Latinas within STEM disciplines within institutions, across industry, and with investments at the state and national levels.

What role does policy play? Public policy is about scale. How do we take the high-impact practices and efforts that effectively serve the small group of Latinas studying and researching in STEM and scale these efforts to serve more Latinas effectively? How can we reinforce their resilience and success while developing the models and strategies to develop interest among more Latinas, cultivate and match with mentors, continue to grow awareness of the strengths and needs of these scholars, and evolve the work needed in STEM?

We have to further accelerate the success of our Latinas in STEM and encourage, support, retain, and graduate more of them. Ultimately, we have to ensure that there is no excuse for not taking action to increase the success of Latinas in STEM. The data are clear—the opportunity to address the recruitment, retention, transfer, financial support, mentoring through faculty, and completion is significant. And we must accelerate Latina student success in STEM, not just increase it. We do not want the closing of gaps in progress and attainment to be the result of a decreased investment in other students. This is not a zero-sum game. While all are improving their educational attainment, we need to accelerate Latina success in STEM to close the gaps with other groups and meet the opportunity we have to increase Latina representation. Authors in this book review theories and data on Latinas in STEM that can inform policy and a tactical plan at scale at institutional, state, and national levels. This review can help to frame a critical understanding of the past, current, and potential asset-based future of Latinas in STEM, and the field overall.

I believe the work we are doing today—informed and guided by the efforts of so many Latinas and Latinos, allies, and supporters who have helped us get to this point in our success—has the potential to accelerate the resilience and success of many more Latinas in STEM. This is our time, and this is happening on our watch. We have a responsibility to create the knowledge, raise our voices, make the space for others, invest in their success, and ensure that others have the opportunity to contribute meaningfully in our collective success.

Deborah A. Santiago, Co-Founder and CEO,
Excelencia in Education

# Notes

1. Snyder, T. D., de Brey, C., & Dillow, S. A. (2020). *Digest of education statistics 2019*. Table 306.10. Washington, DC: National Center for Education Statistics, Institute of Education Sciences, U.S. Department of Education. Retrieved February 11, 2020, from https://nces.ed.gov/programs/digest/2019menu_tables.asp
2. Snyder, T.D., de Brey, C., & Dillow, S.A. (2020). *Digest of Education Statistics 2019*. Table 306.10. Washington, DC: National Center for Education Statistics, Institute of Education Sciences, U.S. Department of Education. Retrieved February 11, 2020, from https://nces.ed.gov/programs/digest/2019menu_tables.asp
3. Snyder, T.D., de Brey, C., & Dillow, S.A. (2020). *Digest of Education Statistics 2019*. Table 318.10, 321, 322, 323, 324. Washington, DC: National Center for Education Statistics, Institute of Education Sciences, U.S. Department of Education. Retrieved February 11, 2020, from https://nces.ed.gov/programs/digest/2019menu_tables.asp
4. Snyder, T.D., de Brey, C., & Dillow, S.A. (2019). *Digest of Education Statistics 2018* (NCES 2020–009). Table 315.20. Washington, DC: National Center for Education Statistics, Institute of Education Sciences, U.S. Department of Education.
5. Snyder, T.D., de Brey, C., & Dillow, S.A. (2020). *Digest of Education Statistics 2019*. Table 318.45. Washington, DC: National Center for Education Statistics, Institute of Education Sciences, U.S. Department of Education. Retrieved February 11, 2020, from https://nces.ed.gov/programs/digest/2019menu_tables.asp
6. Snyder, T.D., de Brey, C., & Dillow, S.A. (2020). *Digest of Education Statistics 2019*. Table 318.10, 321, 322, 323, 324. Washington, DC: National Center for Education Statistics, Institute of Education Sciences, U.S. Department of Education. Retrieved February 11, 2020, from https://nces.ed.gov/programs/digest/2019menu_tables.asp

# Introduction

## An Asset-Based Approach to Advancing Latina Students in STEM: Increasing Resilience, Participation, and Success

*Elsa M. Gonzalez and Miranda Wilson*

In the midst of our collaborative effort to finish this edited volume, the world was rattled by the coronavirus pandemic. Governments across the globe called for solidarity and empowered the public to unite through resilience. This included calls to face the challenge, be strong, make the best of the situation, be successful, not surrender, and persevere. For many of us in academia, these calls sound familiar—a natural response to adversity. The familiar call for resilience resonates and is recognized in our own Latina/o community. This is the response that we want to provide as scholars who see resilience every day in our colleagues, our students, our participants, and our communities. As we sat and planned this edited volume, we agreed on the importance of understanding Latina resilience not only as a scholarly and research endeavor but as a matter of introducing it and making it clear to our Latino community that Latina resilience *is* the response. That response came from a community cultural wealth (Yosso, 2005) with which Latina is identified. However, it is also important to understand and embrace the communal response so that we may make it ours and incorporate it in our actions and decisions as we move through different places and spaces. A year ago, as we keep seeing now, our world, our realities, seemed so different. But were we different? Or were we just as resilient then as now? Maybe external to a global pandemic, we did not see it or recognize it. It is likely, though, we did not actively ask our students to take the elements that form a resilient response everywhere they go, elements they will need as they respond to life's challenges. Now, however, we are doing it. As a natural response to the challenges of this pandemic, we are showing resilience. We are, perhaps, learning that we have been showing it all along.

This book, *An Asset-Based Approach to Advancing Latina Students in STEM: Increasing Resilience, Participation, and Success*, is the result of a continuous conversation among Latinx colleagues who find support and collegiality in the collaborative work that *colegas Latinos* provide to find a space to work together, exchange ideas, and present our research. In this conversation, a recurring question arises: How resilient are Latina students, faculty, administrators, and scholars? More directly, are we resilient? How resilient are

Latinas who study and research in science, technology, engineering, and mathematics (STEM) fields? As the research and work in this edited volume indicate, yes, Latinas are resilient; we claim it and we prove it. Furthermore, we as researchers recognize the importance of understanding why, how, and what makes us resilient in order to apply it as a cultural response to our work, our professions, and our communities.

By way of introducing the chapters in this book, we want to take this opportunity to reflect on a model of Latina resilience (Gonzalez, 2020). This model acknowledges diverse contexts through life and appreciates that Latina resilience is a way of thinking, acting, reacting, and living. Responding to the model, we propose that as scholars we should create or promote the conditions that Latinas need to execute resilience in the present and in the future; which, although those conditions exist, they are often not recognized in one's self or by the people around them. The model presents a contextual approach that goes beyond college into the workplace to confront the challenges that continue as the Latina population introduces itself in more and diverse spaces. Considering the model, we recognize in this book that Latinas need to be aware of exercising resilience across different STEM fields contexts.

A priority exists for Latina students to succeed in STEM fields, and resilience can ensure a successful outcome (Gonzalez, Contreras Aguirre, & Myers, 2020). Having community, family, and institutional support are foundational for Latina students to complete their STEM degrees. Moreover, persistence in STEM fields increase by 14.87% when students join an organization. Working together in a social context as part of a collaboration is a strong motivating factor for women (Grasgreen, 2013). There is a strong connection between Latina students and the work they do. A sense of belonging in college has been determined to be an important factor in student persistence and success during and after college; a sense of belonging can promote access to more student resources/services and mental health (Gopalan & Brady, 2020); however, Latina students need to be aware of their resilience and the positive effect that it can have. Financial burdens are yet another barrier that Latina students can face because of low socioeconomic status, which negatively impacts both success and completion of a degree in the STEM fields (Rendón, Nora, & Kanagala, 2014). This volume demonstrates that even through adversity, Latina students can enhance their success by cultivating resilience through family, peers, mentors, faculty, and identity development. With this information in hand, change makers—those in positions of power—must act to ensure that Latina students' needs are fostered as cultural capital wealth is considered, so that they may transcend through a resilience response in STEM education and beyond.

In 2020, Latinas in schools, colleges, and the workplace are a new generation, a new group that has been added to the traditional Hispanic student group from 10–20 years ago. Many of them are second- or third-generation citizens of the United States; this new group faces unceasing challenges, including racism and sexism in educational settings and the workplace. At the same time, they are in between spaces in their families, society, and schools; they are in the "borderlands," as Anzaldúa (1987) pointed out, with different and particular experiences, different languages, and different ways of knowing. Many are not first-generation, yet many are still "first-gen" in their families. Latinos see and feel this intersection, this "borderland" space, as a source of resilience. The model, first suggested in a special journal issue focused on understanding Latina/o resilience (Gonzalez, 2020), focuses on the various contexts that these students encompass as the population of Latinas/os at large.

The Latina/o resilience model (Gonzalez, 2020) introduces the resilience element that individuals use in challenging contexts, such as STEM fields, not only as students but also as they move through their professions and life experiences. The model embraces and continues the idea of community cultural wealth (Yosso, 2005) as a model of contextual, lifelong resilience.

In 2005, Yosso's model of community cultural wealth elevated the conversation by moving away from a deficit-based perspective of Students of Color for the first time. Fifteen years later, this Latina/o resilience model looks to extend Yosso's work to a specific group, a new generation of Latina/o individuals.

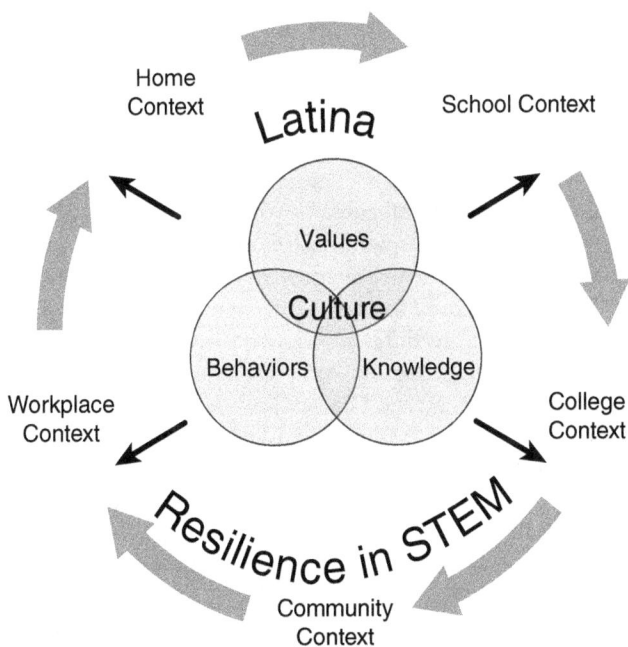

*Figure 0.1* Latina Resilience in Model STEM.

In this book, the model considers the context at home and at school (Chlup et al., 2019), going through college, and then moving to the workforce (including academia) and the community. It reinforces the idea of resilience and encourages the individual to remain mindful of it in all the contexts or spaces where a Latina goes. The resilience model's tenets include the values, behaviors, and knowledge (Rendón, Nora, Bledsoe, & Kanagala, 2019) of the Latina culture as a group; resilience defines the whole group and is present and reflected in different spaces, such as home, school, college, workplace, and community. Latina resilience is supported by the cultural wealth (Yosso, 2005) that Latinas as a group bring to the table. Latinx culture includes assets and ways of knowing (e.g., ethnic consciousness, familial, perseverance, etc.) that add to this cultural wealth and aid in persistence through STEM pathways (Rendón et al., 2019). This model is also supported by Bandura's (2008) work on the development of self. He notes, "People's life pursuits, goals, values, and aspirations influence the form their agency takes and the purposes their efficacy beliefs serve" (p. 28). The Latina/o resilience model (Gonzalez, 2020) demonstrates how values, behaviors, and knowledge converge to encompass Latina/o culture, which is steeped in resilience. And in this volume, we find through the different chapters that resilience then permeates to their agency and beliefs (Bandura, 2008) about Latinas' efficacy in STEM, which can generationally influence Latinas' various and revolving life contexts.

Yosso (2005) introduced the cultural capital wealth (CCW) model to capture the talents, strengths, and experiences that Students of Color bring with them to their college environment. The CCW model represents a framework to understand how Students of Color access and experience college from a strengths-based perspective. Latina resilience does not only address retention and success at college and school; this model addresses success before, during, and after school. The model is an intent of moving and expanding the conversation informed by Yosso's model, wherein the model presents six types of capital: aspirational, familial, social, linguistic, resistant, and navigational. This Latina resilience model proposes to transfer these concepts to a specific group, Latinas. Moreover, we want to make sure that Latina resilience is taken to different contexts as we move the conversation beyond school and college to the workplace. Also, in the Latina resilience model, Latinas' contexts are not linear; they are cyclical and impact one another. More specifically, the home context impacts the school context, which also impacts the college context. All of these contexts converge and are impacted by resilience as Latinas transcend different transitions, including those in and out of STEM classes and careers.

This Latina resilience model considers a continuous acknowledgment of resilience throughout life and clearly identifies as a non-deficit model, asset-based approach. The Latina resilience model focuses the model's concepts specifically on Latinas and their unique experiences. In these chapters, the focused model will identify this group as a new generation of Latinas, a broader group with more complexities—a group where race, gender, and a "borderland" space intersect, and where the model becomes particularly important in STEM areas, still a White-male–dominated space in universities and in the workplace.

## Contribution of This Edited Volume

The authors of the chapters in this book on understanding Latina resilience address different moments of being successful in the present and in the future. Each of the collaborators in this issue provide insights for different contexts.

As a collective, the chapters contribute with components that address Latinas' resilience from various perspectives and unite these components through similar themes and trends. These topics are framed by the crisis of the lack of Latinas throughout STEM pathways from college through the workforce. The edited volume begins with "Latinas in STEM: A Review of the Literature Using a Psychosociocultural Lens." This first chapter compiles literature from a psychosociocultural model perspective based on the work of Gloria and Rodriguez (2000). The psychosociocultural model is used as a standpoint to understand how Latinas persevere in STEM during college. This review highlights how aspects like family, the self, and peers can influence resilience.

Chapter 2, "Developing a Conceptual Framework for Computing Identity Development for Latina Undergraduate Students," builds on the first chapter by providing a framework to understand the Latina student experience in STEM. This chapter specifically investigates Latina students in computing to add a nuanced approach to understanding intersectionality and its impact on identity development. Similar to Chapter 1, this chapter also emphasizes the importance of asset-based perspectives to the development of identity and resilience for Latinas attempting to succeed in a major that traditionally includes White male students and excludes racial and gendered others.

"The Pathway to the PhD: Latinas as STEM Doctorates From 1975–2010," Chapter 3 of the edited volume, approaches STEM resilience by historically analyzing how Latina students have progressed in their attainment of doctoral degrees. The authors of this chapter posit that successful completion of STEM doctorates can be used as a benchmark to understand the lack of Latinas in STEM faculty and leadership positions. Through the analysis of STEM doctoral earner data in the United States, the authors conclude that Latinas are not earning STEM doctorates at the rates implied by the growth of the Latina population, which could be a detriment to increasing Latinas in academia and other STEM-based workplaces.

To continue the discussion of Latina STEM in different spaces within academia, Chapter 4, " 'Cuida Tu Casa Y Deja La Ajena': Focusing on Retention as a Self-Perpetuating Engine for Recruiting Latina Faculty in

STEM," examines the recruitment and retention of Latina faculty by aiming to enhance departmental climates within STEM fields. The authors analyze Latina faculty within astronomical sciences, chemistry, and physics departments to highlight the importance of creating a *familia* atmosphere, which is necessary for Latina faculty to succeed in STEM departments.

Chapter 5, "How Many Latinas in STEM Benefit From High-Impact Practices? Examining Participation by Social Class and Immigrant Status," builds on the positive implications of high-impact practices (HIPs) in fostering resilience for racially disenfranchised students. This chapter specifically demonstrates how the use of HIPs for Latina student success can differ depending on the student's personal characteristics. Through the use of the Student Experience data from the Research University survey, the chapter highlights how Latina social class and immigrant status can influence how HIPs should be used to optimize Latina success in STEM degrees, as a resilient response.

Latina resilience within the context of the research university is expanded in Chapter 6. Entitled "Empowering Latina STEM Majors at a Public R1 Doctoral Hispanic-Serving Institution in Texas: Strategies for Success," this chapter examines the cultural impact of resilience at a top-tier research Hispanic-serving institution. The Latina students in this study collectively expressed the challenges and triumphs of their racialized and gendered experiences while pursuing STEM degrees within this institution. Their Latina background is centered as a source of strength that allows them to foster resilience and persist in STEM majors.

Like the previous chapter, Chapter 7, "First-Generation Latina Engineering Students' Aspirational Counterstories," features the voices of Latina students but also utilizes counterstory methodologies to combat deficit-based narratives about Latina students in STEM. With a specific focus on engineering, this chapter emphasizes changes that must be made within engineering student support services to align with Latina student success strategies. The stories of first-generation Latina students are analyzed and reflect that the theory of validation should be used within engineering support systems to foster resilience for Latina students in STEM.

In Chapter 8, "Latina Undergraduates in Engineering/Computing Science on the US–Mexico Border: Identity, Social Capital, and Persistence," there is a continued focus on Latina identities to understand resilience in mechanical engineering and computer science majors. This study also incorporates the significance of Latina resilience through social capital as a positive enhancement to Latina success to and through these majors. As is done in several of the chapters, social characteristics (e.g., familial, peer) are named as key resilient mechanisms of success for Latina students in STEM.

In Chapter 9, the focus shifts from collective to individual by highlighting the story of a Mexican-American Latina student pursuing a STEM major. In "'I Learned How to Divide at 25': A Counter-Narrative of How One

Latina's Agency and Resilience Led Her Toward an Engineering Pathway," the authors retell Kitatoi's experiences through multiple intersecting identities as she transforms her perception of self to excel as an engineer.

Chapter 10 expounds on the experiences of Latina students in STEM by examining them as leaders. In "Leadership Through the Lenses of Latinas: Undergraduate College Students in STEM-Related Disciples at Regional HSIs," the authors qualitatively examine how gender and ethnicity impact Latina leadership styles when those students are situated within HSIs. Collectively, their narrative demonstrates that Latinas in STEM exhibit leadership by enhancing the climate within the STEM field for other Latina students, and how this can impact their future.

The edited volume concludes with Chapter 11, " 'There was Something Missing': How Latinas Construct Compartmentalized Identities in STEM." This study investigates how Latinas in STEM majors rely on aspects of their identities to persevere. The findings demonstrate how identity compartmentalization is utilized to foster resilience through mentoring with others who possess the same identity.

## Our Readers

This book on Latina resilience and success in STEM is meant not only for researchers, teachers, faculty, and administrators but also for practitioners and policy makers who affect and effect change. Each will find something that is useful and applicable to their contexts, whether academic or future in the workplace, and they will gain a clearer understanding of this dynamic group.

This edited volume seeks to reveal policy implications in higher education and their possible impact in the workplace. We discuss Latina resilience across the whole spectrum of experiences. Our intent is to push the conversation for Latina students to be successful, not just in the short term but beyond graduation. We want to clearly establish that they can be successful if they are supported to have the tools and mindset to be.

Finally, this book presents implications for equity as an outcome of practicing resilience in different contexts. Currently, Latinas decline in numbers at each level of degree attainment/certification (i.e., bachelor, master, doctoral), which results in a growing disparity in positions of academic leadership within STEM fields. Given the current crisis, the demographic realities, and the present national attention on increasing the number of students in STEM disciplines, particularly underrepresented students, university systems must consider factors that motivate and encourage Latinas on their paths to graduate degrees and eventually successful STEM careers (Ruiz, 2013). Through this work, we encourage Latinas to embrace resilience in many diverse contexts in order to be successful in college and beyond, when entering the workforce and when looking back at the paths they forged for others.

## References

Anzaldúa, G. (1987). *Borderlands/La Frontera: The New Mestiza.* San Francisco, CA: Aunt Lute Book Company.

Bandura, A. (2008). Toward an agentic theory of the self. *Advances in Self Research, 3,* 15–49.

Chlup, D. T., Gonzalez, E. M., Gonzalez, J. E., Aldape, H., Guerra, M., & Lagunas, B. (2019). Latina/o high school students' perceptions and experiences obtaining information about going to college: A qualitative study for understanding. *Journal of Latinos and Education.* https://doi.org/10.1080/15348431.2019.1568878

Gloria, A. M., & Rodriguez, E. R. (2000). Counseling Latino university students: Psychosociocultural issues for consideration. *Journal of Counseling and Development, 78*(2), 145–154.

Gonzalez, E. M. (2020). Foreword: Understanding Latina/o resilience [Special issue]. *International Journal of Qualitative Studies in Education.*

Gonzalez, E. M., Contreras Aguirre, C., & Myers, J. (2020). Persistence of HE Latinas in STEM at a R1 institution in Texas. *Journal of Hispanics in Higher Education.* https://doi.org/10.1177/1538192720918369

Gopalan, M., & Brady, S. T. (2020). College students' sense of belonging: A national perspective. *Educational Researcher, 49*(2), 134–137. https://doi.org/10.3102/0013189X19897622

Grasgreen, A. (2013). *Strategy for women in STEM.* Retrieved from www.insidehighered.com/news/2013/11/20/project-based-learning-could-help-attract-and-retain-women-stem-study-suggests

Rendón, L. I., Nora, A., Bledsoe, R., & Kanagala, V. (2019). *Científicos Latinxs: The untold story of underserved student success in STEM fields of study.* San Antonio, TX: Center for Research and Policy in Education, The University of Texas at San Antonio.

Rendón, L. I., Nora, A., & Kanagala, V. (2014). *Ventajas/assets y conocimientos/knowledge: Leveraging Latin@ strengths to foster student success.* San Antonio, TX: Center for Research and Policy in Education, The University of Texas at San Antonio.

Ruiz, E. C. (2013). Motivating Latina doctoral students in STEM disciplines. *New Directions for Higher Education, 163,* 35–42.

Yosso, T. J. (2005). Whose culture has capital? *Race, Ethnicity and Education, 8*(1), 69–91. https://doi.org/10.1080/1361332052000341006

# Part 1

# Examining Literature, Theory, and Data to Inform Policy

# 1 Latinas in STEM

## A Review of the Literature Using a Psychosociocultural Lens

*Kristan M. Venegas and Araceli Espinoza-Wade*

## Introduction

Latinas' underrepresentation in STEM fields has received attention in research and practitioner-based journals over the last ten years. As shown in other chapters in this text, increasing Latina access to and success in STEM fields is a national imperative. In this chapter, we consider literature that emphasizes the psychological, social, and cultural factors that have been explored within the literature and that inform resilience for Latinas in STEM. We have decided to focus on these factors because they make up core elements of resilience. We emphasize a particular model—Gloria and Rodriguez's (2000) psychosociocultural model, or PSC model, which has widely been used to understand how Latinx college students navigate their college experiences. Rather than provide new data, we synthesize research related to Latinas in STEM over the last ten years using a systematic literature review approach to consider the following question: *How does current literature about resilience for Latina college students in STEM connect to the psychosociocultural model?*

The literature review approach that we describe in the following section serves as our method for analysis. The articles we have identified through a multi-step process of review serve as our data. As a result of this meta-analysis, we are able to synthesize findings and suggest future implications for research and practice. We rely on Butler's classic definition of resilience and work to relate it to the PSC model. Butler defines resilience as "an interactive and systemic phenomenon, the product of a complex relationship of inner strengths and outer help" (1997, p. 26). There are additional forms of resilience that are likely to inform the research that we have reviewed. For example, Reyes (2012) situates Latinas' academic resilience in postsecondary education as a form of both achievement and resistance. Like the other chapters of this text, we interpret resilience as contextual, multifaceted, dynamic, and personal.

## Theoretical Framework: Psychosociocultural (PSC) Model

The psychosociocultural approach includes the exploration of three elements as they occur on college and university campuses. These components include the psychological, social, and cultural dimensions of postsecondary education (Gloria & Rodriguez, 2000). These components are intertwined and at times can overlap based on the context and experiences of the individual. The psychological features of the model include the beliefs, attitudes, and perceptions individuals hold about themselves. The social elements of the model focus on the networks, mentors, peers, role models, and other social connections related to career and campus life. Finally, the cultural piece of the model is inclusive of the values, meaning, and possible validation that individuals might experience. These three components come together as a persistence model that seeks to offer an explanation of how Latinx students—and in this case, Latina students—find their way through their academic major and daily campus life.

Gloria and Rodriguez's (2000) model emerges from the field of counseling. The model is attentive to prior research about student development and acknowledges research showing that all students face adjustment issues as they come into college. However, Gloria and Rodriguez (2000) also make visible the history of discrimination, and the perceptions and realities of socioeconomic differences that Latinx students may face on college campuses, which places them at a disadvantage in comparison to their White peers. Drawing on the work of Sue, Arrendondo, and McDavis (1992), Gloria and Rodriguez question the effects of ethnic identity and acculturation and its possible effects on the mental health and persistence of Latinx college undergraduates. After presenting qualitative data based on interviews with Latinx college students, they outline a dispositional "set of attitudes and beliefs, knowledge, and skills" that counselors should use when counseling Latinx students (Gloria & Rodriguez, 2000, p. 150). These dispositions should be used to help counselors consider their own competencies related to students' dispositions, as well as the "university environment, ethnic identity, acculturation, and social identity" while working with Latinx students (p. 152). Studies that explore the PSC model quantitatively (Aguinaga & Gloria, 2015; Gloria, Castellanos, & Herrera, 2016) and qualitatively (Moreno & Banuelos, 2013; Sanchez, 2011) continue to support that the concerns of Latinx students noted in the additional study continue, and that attention to counseling and student affairs practices that encompass these supporting practices are valuable.

For example, Gloria, Castellanos, and Orozco (2005) center coping strategies that are fundamental to the PSC model and college persistence in their study of 98 Latinas pursuing graduate-level training. Their study included Latinas who differed by educational goals, characteristics, and immigration generations. Their core findings suggested that positive action as a coping response and cultural congruity were consistent predictors of well-being and degree completion. Cultural congruity can be described as understanding how to support the psychosociocultural needs of these students. The authors were intentional in pointing out that these positive signifiers of well-being pushed back against stereotypes about Latinas, how they value education, and perceptions about their ability to mitigate barriers to their educational plans.

Castellanos and Gloria's (2007) research using the PSC model asserted that three elements should come together to form strategies for success. Relying on qualitative interviews with Latinx students on a predominately White college campus, they suggest that there are numerous opportunities to improve the college experience for Latinx students. Some of the psychological elements of success would include: "receiving a monthly care package from home . . . talking with family to provide daily updates . . . (and) having faculty ask and be concerned about one's well-being" (p. 386). Social elements of success could be: "attending monthly Latina/o based student organization meetings . . . weekly meetings with a faculty member to discuss educational progress . . . greeting Latina/o peers on campus between classes" (p. 386). Finally, cultural elements might be: "engaging in monthly community projects that address Latina/o issues . . . talking about family with a faculty member over coffee or lunch . . . (and) fluidly moving between an ethnic-specific student group and predominantly white classroom" (p. 387). Through these recommendations for success, we can visualize an achievable set of practices that would shift both policies and cultural practices related to retention on a college campus.

As shown in Castellanos and Gloria (2007), the PSC model is intentional about identifying factors, practices, and policies that influence Latina college persistence. The psychological elements include self-esteem and self-efficacy. The social elements can include family, mentors, and peers. Cultural elements include one's connection to ethnic identity and cultural congruity or incongruity that one experiences in the university context (Castellanos & Gloria, 2007). The PSC model can be used to approach persistence goals by examining the experiences of the individual and intentionally connecting them to practices to improve the campus climate.

## Methods: Developing a Review of the Literature

A literature review requires systematic steps in identifying literature (Galvan & Galvan, 2017). Some of these steps including using key search terms and multiple search engines. We chose to use Google Scholar, JSTOR, and PsycInfo. We selected Google Scholar as a search engine because we knew that we would be able to include references that might move beyond the JSTOR requirements, and that it is globally inclusive of educational and psychological research. According to its website, JSTOR catalogues journals for 75 disciplines and includes more than 12 million journal articles. We used the search terms "Latinas", "computer engineering", "computer science", "college", "psychological factors", "social factors", and "cultural factors". Using the search terms "Latinas and computer engineering and resilience" yielded over 15,900 results between 2010–2020. When we narrowed to "Latinas and computer science and resilience" the search shifted to

*Table 1.1* Summary of First Set of Search Terms and Number of Responses.

| Search Terms | Google | JSTOR | PsycInfo |
| --- | --- | --- | --- |
| Latinas, computer engineering, resilience | 15,900 | 5 | 0 |
| Latinas, computer science and resilience | 14,300 | 19 | 1 |
| Latinas, computer science and engineering and psychological factors in college and resilience | 18,000 | 32 | 321 |
| Latinas, computer science and engineering, and social factors in college and resilience | 17,500 | 48 | 321 |
| Latinas, computer science and engineering, and cultural factors in college and resilience | 17,300 | 42 | 293 |
| Latinas, computer science and engineering, and psychosocial cultural factors in college and resilience | 16,700 | 55 | 367 |

14,300 results. Our next step was to complete a search based using the term "Latinas and computer science and engineering" and each of the PSC factors. Table 1.1 shows the results of our search. PsycInfo includes more than 2,500 journals, the focus of which are inclusive of psychological topics in both behavioral and social sciences.

While at first glance it might seem that there is a promising and broad discussion of Latinas, resilience, and psychosociocultural factors in college, a deeper view of this literature and search engines revealed that there were up to 55 articles in JSTOR and 367 articles in PsycInfo that were relevant to addressing our guiding research question. Again, relying on Galvan and Galvan's (2017) recommended process, we reviewed the JSTOR and PsycInfo articles to further narrow to 20 articles in JSTOR and 55 articles in PsycInfo. For Google, we narrowed our search to include only the terms noted in Table 1.2 in the title and narrowed to journal articles. Table 1.2 summarizes the next phase of review. In this phase, we narrowed our search as follows.

Once we narrowed to this set of articles, we reviewed them for overlap and content. We realized that these studies, while inclusive of Latinas, tended to see Latinas as one racial/ethnic and gendered category in an overall study of historically underrepresented racial/ethnic groups. In some studies, Latinas were part of a study that aggregated Latinx/as as one category of individuals so that gender differences might not be fully explored. This lack of gender breakdown meant that we could not separate which findings might be more relevant to the specific Latina STEM experience. In other studies, Latinas might be categorized with other women of color—which, again, makes it difficult to isolate and honor the unique experiences of Latinas. While these studies undoubtedly have value within academia, our goal in this chapter is to center the experiences of Latinas in STEM. The result of this sorting led us to our third and final round of review. During this round, we narrowed

Table 1.2 Summary of Second Set of Search Terms and Number of Responses.

| Search Engine | Search Terms | Remaining Relevant Articles |
| --- | --- | --- |
| JSTOR | Latinas, computer science and engineering, and psychosocial cultural factors in college and resilience | 20 |
| PsycInfo | Latinas, computer science and engineering, and psychosocial cultural factors in college and resilience | 55 |
| Google | Latina and STEM; Chicana and STEM via advanced search in title | 22 |

our scope by reviewing each piece of literature to ensure that each article met the following criteria:

1. Must focus on Latinas and/or Chicanas in STEM in the United States.
2. Must have a peer review process. Dissertations and thesis studies were included because they have a peer review process. Policy or advocacy papers without evidence of a peer review were removed.

There are two exceptions that we applied to these criteria. First, we included Espinoza's (2013) study; although it includes both Latinx and Latina students, we decided to include it because it utilizes the PSC model. Second, we decided to include additional references from Rodriguez (2018), as well as a number of chapters within this text because although they did not appear in the search term process, as authors, we knew of these studies and knew that they met the criteria for study. We were able to identify this need for inclusion through our close reading of the final set of articles. Even with using this strict identification process, it is likely that there are rigorous studies that were not included here.

As part of following a step-by-step literature review process, we needed to create a set of criteria through which we could identify key studies (Galvan & Galvan, 2017). Table 1.3 provides a summary of key studies. It includes the author and year published, the study population, theoretical framework, research methods, and a short summary of core findings.

Of the 21 studies noted here, 16 are qualitative, four are mixed method, and one relies on structural equation modeling. Family support was the most often noted finding (21), followed by self/gender (16), campus climate or environment (11), faculty and mentors (11), and peers (10) and orientation towards career (4). It is important to note that academic preparation did not emerge as a challenge or support within these studies. That is not to say that Latina students in the studies might not experience academic challenges, but perhaps the study respondents found other aspects of their STEM preparation and/or campus social experiences to be more challenging parts of their academic journeys.

*Table 1.3* Summary of Key Studies Related to Latinas, STEM, and the Psychosociocultural Model.

| Authors | Theoretical Framework | Research Methods | Core Findings |
|---|---|---|---|
| Aguirre and Banda (2019) | Relational Cultural Theory | Case study on two campuses | Formal female mentors and information family mentors made the difference towards success |
| Arroyo (2017) | Critical analysis | Testimonio | Mentorships, undergraduate research programs, family support, peer support, and career awareness toward success |
| Banda and Flowers (2018) | Critical consciousness | Counter-narratives | Core themes: Desire to relate to other minoritized populations/women and experiencing sexism and discrediting one's engineering competency |
| Bello (2018) | | Multiple case study | Incremental mindset was valuable for success |
| Castellanos (2018) | PSC model and social cognitive career theory | Structural equation modeling | Campus climate, academic involvement, faculty support played a role in Latinas' career decision-making process. Latinas with higher socioeconomic status, faculty support, and academic involvement more likely for STEM career |
| Cantú (2012) | Funds of knowledge | Testimonio | Roles of parents, teachers, extended family, and community; the impact of the intersection of gender, race, and class |
| Espinoza (2013) | Psychosociocultural model | Semi-structured interviews | Wanting to do well for themselves and families enhanced participants' desires to persist; accessibility of professors determined whether or not participants felt validated; participants differentiated between the climate on campus and the climate within the school of engineering |
| Esquinca, Villa, Hampton, Ceberio, and Wandermurem (2015) | Sociocultural theory | Semi-structured interviews | Overcoming adversity depended on many factors, such as the role of peer groups, family, and professors as well as spaces that accommodate group interaction |

| Authors | Theoretical Framework | Research Methods | Core Findings |
|---|---|---|---|
| Garcia, Rincón, and Hinojosa (2020) | Resilience and identity theories | Qualitative interviews derived from a mixed-method study | Resilience is achieved through the construction of compartmentalized identities within STEM, which was leveraged to establish same-identity mentoring relationships |
| Gonzalez, Molina, and Turner (2020) | Resilience theories | Exploratory/ descriptive qualitative case study | Desire to succeed and move up in life and well as mentor, be mentored, and persevere |
| Hiles (2015) | Not clearly stated | Mixed method | Alienation and lack of support was not statistically present, but this and similar issues were addressed strongly in qualitative responses |
| Longoria (2013) | Community Cultural Wealth | Semi-structured interviews | Emphasized the value of community cultural wealth, academic preparation, and impact of socioeconomic circumstances and sexism |
| Martínez et al. (2019) | Contextual mitigating factors | Case study | Explained incidents of racism, sexism, and STEM assessments that served as potential gatekeepers |
| Mein, Muciño, Guerra, and Herrera-Rocha (2020) | Identities in practice | Case study | Four key types of social resources were valuable, including family, friendships, peers, and mentor-based resources |
| Mercédez (2015) | Community cultural wealth and science identity | Mixed method | Survey results showed that performance, competence, and recognition were most valuable<br>Interview results showed that family of origin and community on campus were valuable, and that chilly climates were present |
| Rodriguez, Cunningham, and Jordan (2019) | Phenomenological approach | Semi-structured interviews | Focused on self-identity within STEM, internal and external recognition from peers and faculty were valuable in developing this identity |
| Rodriguez, Doran, Sissel, and Estes (2019). | Phenomenological approach | Semi-structured interviews | Professional identities are developed through interactions with family members, identity-based organizations, and in relation to their other identities |

*(Continued)*

*Table 1.3* (Continued)

| Authors | Theoretical Framework | Research Methods | Core Findings |
|---|---|---|---|
| Rodriguez, Lu, and Ramirez (2020) | Relies on previous related work to present a cultural framework | Review of scholars' current body of work | Considers how Latinas in STEM develop their identity with attention to community cultural wealth, funds of identity, and sense-making |
| Rodriguez, Pilcher, and Garcia-Tellez (2019) | Phenomenological approach | Semi-structured interviews | Interdependence and attachment to family members supports the STEM journey; family members in and outside of STEM field influenced identity development; and complications emerged from intertwining STEM, family, and "good daughter" identities |
| Tello (2015) | Constructivist Grounded Theory and Psychosocial model | Semi-structured interviews | Campus involvement, relational support, and cultural influences aided in their college persistence and graduation; campus climate did not |
| Verdín (2020) | Agency, environmental factors, and disciplinary role identities | Counter-narrative | The individual in this study navigated three environmental structures: imposed, selected, and constructed |

## Analysis: Viewing the Research Through a PSC Lens

The PSC model includes three areas of influence—psychological, social, and cultural—as a part of understanding Latina college students' persistence. We sought to consider resilience and retention using the PSC model and the following research question: *How does current literature about resilience for Latina college students in STEM connect to the psychosociocultural model?* After identifying the 16 most salient articles, we found varying results in terms of which constructs of the model were most present.

In terms of the psychological element, which includes self, 16 of 21 studies emphasized one's sense of self-efficacy and self-esteem as highly important. S. Rodriguez's work—published throughout 2019 with multiple co-authors and within this text, in partnership with Lu and Ramirez—as well as Arroyo (2017)'s work, focus on the need to bring together one's multiple types of identities to create a successful and positive sense of self. Rodriguez and colleagues' research suggests that the development of an integrated identity would increase STEM completion for Latinas. Arroyo's work, which combines critical analysis and testimonio, reinforces these findings of the value of a constructive sense of self.

With regard to the social element of the PSC model, we again see a number of findings that overlap. Family influence was noted as being important in all 21 studies that we reviewed. In each study, family was highlighted as a source of strength (Hiles, 2015), connection (Longoria, 2013; Martínez et al. (2019), and understanding (Tello, 2015), and as part of the structure of one's life (Verdín, 2020). Of course, many studies acknowledged that especially in the case of a first-generation student, the family might not have a detailed understanding of the student experience, but students' knowledge that their families, including extended family, supported them had a motivating effect (Espinoza, 2013; Mein et al., 2020; Verdín, 2020).

Eleven studies noted affirmative interactions with faculty and mentors. For example, Cantú's work (2012) showcased the transformational effects of the faculty or mentor who takes notice in the STEM promise of the individual. She notes that this mentoring role can begin before college and carry into postsecondary experiences. Gonzalez, Molina, and Turner's work (2020), as well as Garcia et al.'s (2020) work, consider the value of mentors as essential to resilience. This seems especially salient in the work of Garcia et al., which suggests that mentors might serve as a bridge throughout an otherwise compartmentalized academic experience. Other studies that emphasize the value of faculty and mentors suggested that even the mere presence of faculty, especially Latina faculty, as visual role models held a positive effect (Aguirre & Banda, 2019; Bello, 2018; Castellanos, 2018).

Peers, which can be considered part of the social construct in the PSC model as well as part of the cultural construct, emerged as an influential group within the context of these studies. While some studies clearly called out male-dominant culture within STEM departments (Arroyo, 2017; Bello, 2018; Cantú, 2012; Mercédez, 2015), they also noted that Latinas were able to find community with other Latinas outside of their majors, which helped to mitigate feelings of isolation within the major. Mein et al.'s findings (2020) suggested that peers are a critical part of understanding and developing one's identity in practice. Rodriguez et al. (2019) and Rodriguez et al. (2019) highlighted the value of student organizations as part of the social element. Peer organizations served as connections between the social and cultural elements of the PSC model.

Ethnic identity and cultural congruity are part of the cultural element of the model. As Rodriguez et al.'s (2019) study highlights, it was the connection with identity-based student organizations that made a difference for the students in their study. Espinoza's (2013) qualitative study of first-generation Latinxs in STEM fields came to a similar conclusion, as does Esquinca et al.'s (2015) qualitative study of Latina undergraduates in computer science and engineering. Esquinca et al.'s (2015) findings not only supported the value of identity-based organization; they also advocated for consistent spaces that supported group interactions like these.

There were 11 studies that found that campus climate, or environmental factors related to the campus or academic major experiences, had an effect on resilience or persistence. These studies shared mixed results. Tello (2015), Banda and Flowers (2018), and Martínez et al. (2019) identified campus climates as spaces that were sometimes neutral and other times chilly. Verdín's work suggested that these environments are part of an individual's selected or imposed environment. While a Latina STEM student might choose to attend a particular campus or select a major, they would have little control over the environment overall. Other qualitative and mixed-methods studies within this literature review found similar results. These studies reported more challenges to resilience as part of the external environment, rather than an internal struggle.

The one outlying finding from these studies was the value that Latinas place on careers and career orientations. Four studies—Arroyo (2017), Castellanos (2018), Gonzalez et al. (2020), and Rodriguez et al. (2019)—mentioned career goals or aspirations as a main driver for the women in their study. These findings could be a result of the kinds of questions that were asked in these qualitative and quantitative studies. In each case, the aspiration toward career was noted as a positive effect.

## Conclusions and Implications

The goal of this chapter was to consider how current literature about resilience for Latina students in STEM might be connected to the three elements of the psychosociocultural model. By using a systematic process for identifying literature, we were able to narrow a very broad search to 21 core pieces that related to Latinas, STEM, and resilience. We then analyzed them to identify their main findings. We found that overall, the characteristics that define the PSC model were aligned with the main outcomes of these studies.

One could argue that self-esteem and sense of self is a vital part of achieving any goal. The "P" or psychological element of the PSC model focuses on self. Over half the studies noted here included a finding that emphasized the sense of self and self-esteem. The work presented by Rodriguez and co-authors (2019, 2020) consistently made the case for the development of a connected sense of self that is inclusive of what she calls a "STEM identity". While not as explicitly named in other studies, the need for this kind of integration of oneself into one's academic or career plans was present and deserved further attention as a matter of research and practice. The PSC model, which emerges from a counseling approach, is well suited to address the academic or career plans of Latinas in STEM fields.

The "S" or social aspect of the model was most often mentioned; the support of family, faculty, and mentors were vital to Latina STEM student success. In terms of family, recommendations included finding ways to authentically involve families by increasing their understanding of what a STEM major and career might look like in terms of time and other expectations. Additional recommendations suggested that families be invited to campus to perceive the university as an integrated part of the family unit's life. It seemed that family exhibited the sense of caring that helped Latina STEM students in their resilience. To further refine this support, family members would benefit from having a better understanding of how to demonstrate caring in ways that would support students. Conversely, faculty and mentors had great influence over Latina STEM students. They were encouraged to continue to show their sense of caring as a means of helping build students' resilience and persistence. There is an alignment between these recommendations and Castellanos and Gloria's (2007) student success practices.

The "C" or cultural portion of the model overlaps across findings related to peers and overall cultural congruity. Peers were seen as helpful overall, especially when connected to student organizations. These student organizations were valuable whether tied to an ethnic identity or not; the important benefit from exposure to student organizations was the authentic connection that individuals felt to the group. The campus climate was also viewed within these mixed results. Once again, there is alignment between Castellanos and Gloria's (2007) student success practices and the kinds of recommendations that are found in these research studies.

This chapter presented an opportunity to consider current research and connect it to a counseling model that has been shown to improve educational outcomes for Latinx college students. Based on a review of this literature, there are connections between the PSC model, Latina STEM students, and their demonstrated performance and values related to resilience. As noted elsewhere in this chapter, the students did not express excessive stress related to their academic performance. Further research that focuses specifically on academic resilience and its relationship with academic stress for Latinas in STEM majors would be valuable. While the findings in this study revealed that other interpersonal relationships with peers, family, faculty, and mentors were central to helping them find success, it would be valuable to delve into how romantic relationships might also impact Latina and STEM degree completion. Through our review of the literature, we noted that there were a few studies that considered the effect of romantic relationships and children on women in STEM careers. It seems that more research is needed to understand how these relationships might impact Latinas specifically as part of degree completion and career preparation. Finally, PSC as a model of practice does include viable solutions that can be implemented to boost Latina STEM degree persistence. It is our hope that practitioners, faculty, mentors, and all those who support Latinas in STEM degrees seek to understand how to support the psychosociocultural needs of these students.

# References

Aguinaga, A., & Gloria, A. M. (2015). The effects of generational status and university environment on Latina/o undergraduates' persistence decisions. *Journal of Diversity in Higher Education, 8*(1), 15–29.

Aguirre, H. C. C., & Banda, R. M. (2019). Importance of mentoring for Latina college students pursuing STEM degrees at HSIs. *Crossing Borders/Crossing Boundaries*, 111–127.

Arroyo, J. L. (2017). *Latina women in STEM: A critical analysis of Ph.D. students' experiences* (Doctoral dissertation). Retrieved from ProQuest Dissertations and Theses database. (UMI No. 10602614).

Banda, R. M., & Flowers, A. M. (2018). Critical qualitative research as a means to advocate for Latinas in STEM. *International Journal of Qualitative Studies in Education, 31*(8), 769–783.

Bello, B. (2018). *Exploring Latina and Hispanic female students' sense of belonging in STEM majors following a belonging intervention* (Doctoral dissertation). Retrieved from ProQuest Dissertations and Theses database. (UMI No. 10935424).

Butler, K. (1997). The anatomy of resilience. *Family Therapy Networker, 21*, 22–31.

Cantú, N. (2012). Getting there cuando no hay camino (when there is no path): Paths to discovery testimonios by Chicanas in STEM. *Equity & Excellence in Education, 45*(3), 472–487.

Castellanos, J., & Gloria, A. M. (2007). Research considerations and theoretical application for best practices in higher education: Latina/os achieving success. *Journal of Hispanic Higher Education, 6*(4), 378–396.

Castellanos, M. (2018). Examining Latinas' STEM career decision-making process: A psychosociocultural approach. *The Journal of Higher Education, 89*(4), 527–552.

Espinoza, A. (2013). The college experiences of first-generation college Latino students in engineering. *Journal of Latino/Latin American Studies, 5*(2), 71–84.

Esquinca, A., Villa, E. Q., Hampton, E., Ceberio, M., & Wandermurem, L. (2015, October). Latinas' resilience and persistence in computer science and engineering: Preliminary findings of a qualitative study examining identity and agency. In *2015 IEEE Frontiers in Education Conference (FIE)* (pp. 1–4). IEEE.

Galvan, J. L., & Galvan, M. C. (2017). *Writing literature reviews: A guide for students of the social and behavioral sciences.* New York, NY: Routledge.

Garcia, A., Rincón, B., & Hinojosa, J. K. (2020). "There was something missing": How Latinas construct compartmentalized identities in STEM. In E. M. Gonzalez, F. Fernandez, & M. Wilson (Eds.), *Latina women studying and researching in STEM: An asset-based approach to increasing resilience and retention.* New York, NY: Routledge.

Gloria, A. M., Castellanos, J., & Herrera, N. (2016). The reliability and validity of the cultural congruity and university environment scales with Chicana/o community college students. *Community College Journal of Research and Practice, 40*(5), 426–438.

Gloria, A. M., Castellanos, J., & Orozco, V. (2005). Perceived educational barriers, cultural fit, coping responses, and psychological well-being of Latina undergraduates. *Hispanic Journal of Behavioral Sciences, 27*(2), 161–183.

Gloria, A. M., & Rodriguez, E. R. (2000). Counseling Latino university students: Psychosociocultural issues for consideration. *Journal of Counseling and Development, 78*(2), 145–154.

Gonzalez, E. M., Molina, M., & Turner, S. C. (2020). Empowering Latina STEM majors at a public R1 doctoral university and Hispanic-serving institution in Texas:

Strategies for success. In E. M. Gonzalez, F. Fernandez, & M. Wilson (Eds.), *Latina women studying and researching in STEM: An asset-based approach to increasing resilience and retention.* New York, NY: Routledge.

Hiles, H. R. (2015). "Our experiences are not unique": An exploratory study of common motivators and inhibitors for Latinas in STEM fields. *iConference 2015 Proceedings.* JSTOR. Retrieved January 1, 2020, from https://about.jstor.org/

Longoria, E. (2013). *Success stems from diversity: The value of Latinas in STEM* (Master's thesis). California State University, Northridge.

Martínez, A. J. G., Pitts, W., de Robles, S. L. R., Brkich, K. L. M., Bustos, B. F., & Claeys, L. (2019). Discerning contextual complexities in STEM career pathways: Insights from successful Latinas. *Cultural Studies of Science Education, 14*(4), 1079–1103.

Mein, E., Muciño Guerra, H., & Herrera-Rocha, L. (2020). Latina undergraduates in engineering/computer science on the US-Mexico border: Identity, social capital, and persistence. In E. M. Gonzalez, F. Fernandez, & M. Wilson (Eds.), *Latina women studying and researching in STEM: An asset-based approach to increasing resilience and retention.* New York, NY: Routledge.

Mercédez, C. D. L. (2015). *Patterns of persistence of Latinas in science, technology, engineering, and mathematics (STEM) degree programs: A mixed method study* (Doctoral dissertation). University of Texas, Austin.

Moreno, D., & Banuelos, S. M. S. (2013). The influence of Latina/o Greek sorority and fraternity involvement on Latina/o college student transition and success. *Journal of Latino/Latin American Studies, 5*(2), 113–125.

PsycInfo. Retrieved January 1, 2020, from www.apa.org/pubs/databases/psycinfo/?tab=3

Reyes, R. A. (2012). *Proving them wrong: Academically resilient first-generation Latinas in college* (Doctoral dissertation). Retrieved from ProQuest Dissertations and Theses database. (UMI No. 3521375).

Rodriguez, S. L. (2018). *Women of color: Architects for systemic change in tomorrow's higher education landscape.* Washington, DC: Annual National Association of Student Personnel Administrators (NASPA) Knowledge Community Publication. NASPA.

Rodriguez, S. L., Cunningham, K., & Jordan, A. (2019). STEM identity development for Latinas: The role of self-and outside recognition. *Journal of Hispanic Higher Education, 18*(3), 254–272.

Rodriguez, S. L., Doran, E. E., Sissel, M., & Estes, N. (2019). Becoming la ingeniera: Examining the engineering identity development of undergraduate Latina students. *Journal of Latinos and Education,* 1–20.

Rodriguez, S. L., Lu, C., & Ramirez, D. (2020). Developing a conceptual framework for computing identity development for Latina undergraduate students. In E. M. Gonzalez, F. Fernandez, & M. Wilson (Eds.), *Latina women studying and researching in STEM: An asset-based approach to increasing resilience and retention.* New York, NY: Routledge.

Rodriguez, S. L., Pilcher, A., & Garcia-Tellez, N. (2019). The influence of familismo on Latina student STEM identity development. *Journal of Latinos and Education,* 1–13.

Sanchez, S. M. (2011). *First-generation Latino males in Latino fraternities at a predominately White institution: Psychological, social, and cultural college experiences.* (Doctoral dissertation). Retrieved from ProQuest Dissertations and Theses database. (UMI No. 3478002).

Sue, D. W., Arrendondo, P., & McDavis, R. J. (1992). The conceptual framework. *Journal of Consulting and Development, 70*(4), 477–486.

Tello, A. M. (2015). *The psychosocial experiences of Latina first-generation college graduates who received financial and cultural capital support: A constructivist grounded theory.* (Doctoral dissertation). Retrieved from ProQuest Dissertations and Theses database. (UMI No. 3702397).

Verdín, D. (2020). "I learned how to divide at 25": A counter-narrative of how one Latina's agency and resilience led her toward an engineering pathway. In E. M. Gonzalez, F. Fernandez, & M. Wilson (Eds.), *Latina women studying and researching in STEM: An asset-based approach to increasing resilience and retention.* New York, NY: Routledge.

# 2 Creating a Conceptual Framework for Computing Identity Development for Latina Undergraduate Students

*Sarah L. Rodriguez, Charles Lu, and Daisy Ramirez*

Computing and computer technology are part of everyday life for people living in the twenty-first century. Just about everything that touches peoples' lives—from the cars we drive, to the movies we watch, to the ways businesses and governments interact with us—is embedded in computing machinery. The field is the primary means by which innovation happens and what allows people to advance as a society. A study by the Pew Research Center (2017) found that four in ten Americans credit computing as the largest contributing factor to improving life in the past 50 years. The same study by the Pew Research Center (2017) found that one in five Americans expect computing to be the largest contributing factor to improving life in the next 50 years. As such, the computing industry is expected to continue growing at a fast pace. Currently, computer science is one of the fastest-growing and highest-paying career paths in the world. It is anticipated that there will be a million more computing jobs than students by the year 2020, which represents a $500 billion national economic opportunity (Lynch, 2016), as well as a personal economic benefit for students, as these careers often offer lucrative salaries.

Although Latina students are entering higher education at greater rates than before, they continue to have disproportionately lower completion rates and career representation in computing than their peers, making up only 2% of all bachelor's degrees earned in computing and 5% of all women employed in computing occupations (National Science Foundation, 2016). There has been relatively no growth in computer science bachelor's degree attainment for Latina students over the past two decades (National Science Foundation, 2019). Although bachelor's degree attainment for Latina students has doubled in other STEM disciplines, such as biological sciences (2.68% to 5.94%), physical sciences (1.87% to 3.33%), and math and statistics (1.57% to 3.51%), computer science is the only discipline that has remained relatively flat (1.75% to 1.79%) (National Science Foundation, 2019).

Ample research has shown that women and racially minoritized people continue to be underrepresented in the computing field (e.g. Sax et al.,

2018; Sax, Zimmerman, Blaney, Toven-Lindsey, & Lehman, 2017; Stout & Blaney, 2018). In fact, while more than half of all bachelor's degrees are awarded to women, only 20% of degrees in computing disciplines are earned by women (National Center for Education Statistics, 2017; National Science Foundation, 2017). Research has also shown that women and people of color often face negative stereotypes about their ability to succeed in computing disciplines (Barker, McDowell, & Kalahar, 2009; Cheryan, Plaut, Davies, & Steele, 2009; Margolis, Estrella, Goode, Holme, & Nao, 2008; Varma, 2010). When women encounter events or situations that reinforce stereotypes of computing as a masculine or male-dominated discipline, many of them question their decision to enter into the field (Lewis, Stout, Pollock, Finkelstein, & Ito, 2016). Additional scholarship has shown that women and people of color pull out of computing disciplines because they do not feel like they belong in the field (Barker et al., 2009).

Furthermore, while women across computing have identified issues related to gender stereotypes and socialized beliefs about who can succeed, Latina students, as women of color, may also endure racialized experiences in which they are marginalized not only by their gender but also by their race and ethnicity. Such hostile environments and marginalization can make it difficult for Latina undergraduate students to develop and maintain a computing identity (Rodriguez & Lehman, 2018). To date, few articles have addressed computing identity, and there are currently no conceptual frameworks that seek to understand computing identity theory and development for Latinas. Over time, scholars have advocated for greater consideration of such intersectionality (e.g. Crenshaw, 1991), yet it is only recently that computing scholars have called for integrating such considerations into computing identity theory. Rodriguez and Lehman's (2018) paper, for example, advocated for a strong need for an enhanced theoretical framework that explored how computing environments interact with and may potentially create obstacles for computing identity development. Further, they argued that understanding how college students develop a computing identity and why some students fail to see themselves as computer scientists may be key to designing effective strategies to recruit more students, particularly more women and underrepresented racially minoritized students, to the computing major and retain them in the field.

This chapter seeks to answer the call for computing identity theory specifically for Latina undergraduate students. To address the need for better computing identity theory, we propose a multi-part conceptual framework for computing identity development for Latina undergraduate students. In previous work, researchers have called for a greater intersectional and theoretical understanding of such identity development processes in order to create more equitable computing learning environments (Rodriguez & Lehman, 2018). This chapter builds upon prior work by contextualizing this call and presenting a framework for identity development specifically for

Latina undergraduate computing students. This framework emphasizes the role that intersectional identities have on elements of computing interest, as well as identity performance and recognition. This framework is informed by prior literature and an ongoing research agenda focused on illuminating how Latina students make meaning of their experiences and develop and sustain computing identities during college. Ultimately, we advocate for utilizing what we know about identity development to support a policy and practice-based holistic approach to understanding how multiple, intersecting identities influence computing identity. We encourage stakeholders to consider how they might encourage resilience through identity development while simultaneously acknowledging the structural oppressions that require Latina students to have to be resilient in computing.

## Proposed Conceptual Framework

We propose a conceptual framework for computing identity development for Latina undergraduate students that consists of four main parts: (1) iterative computing identity development at the individual level; (2) five interweaving systems that define the computing and broader environment; (3) community cultural wealth and funds of identity derived from those environments; and (4) a distinct recognition of intersectionality and multiple forms of oppression these women may experience. As a conceptual framework, this work brings together multiple existing frameworks to explain how Latina undergraduate students make sense of their computing experiences and identities within their various environmental contexts. See Figure 2.1 for a visual representation of this conceptual framework.

### Computing Identity Development at the Individual Level

We start at the individual level at the center of the proposed conceptual framework. At the individual level, we have students coming in with pre-college computing identity experiences. Once in college, they gain computing identity experiences that may cause them to engage in daily identity negotiations and sense-making. The result is a revised computing identity and subsequent college outcomes. In the next section, we describe more about what happens within this individual-level process of computing identity development.

### Borrowing From a Scrum Model

When putting this conceptual framework together, we thought about it almost as an agile, Scrum-like model, similar to that used in software development (Team Linchpin, 2019). Within software development, Scrum is a framework for managing a complex process. We feel as though this will

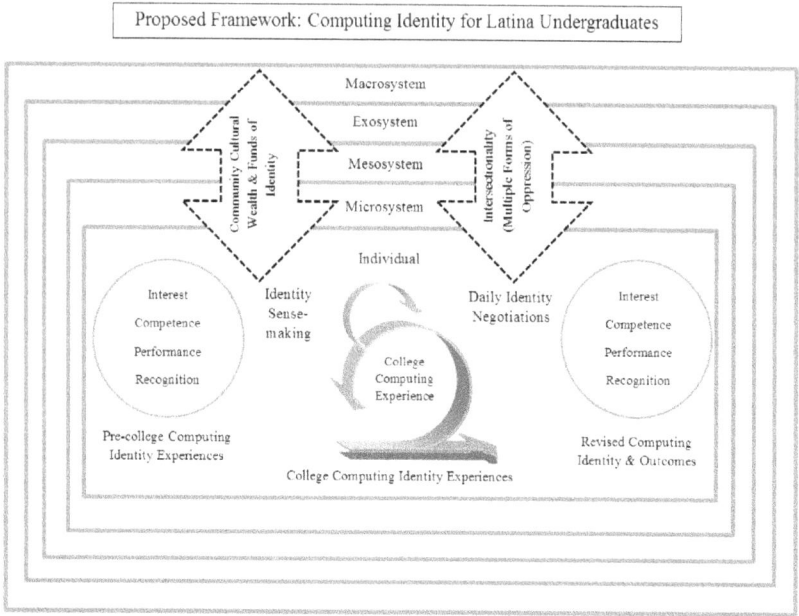

*Figure 2.1* Proposed Framework for Computing Identity Development for Latina Undergraduates.

resonate with computer scientists, as it is a framework that they are already familiar with but applied to identity development. As a process, Scrum emphasizes interaction among stakeholders and attends to reflection and readjustment. It is incremental, iterative, responds well to unpredictability, and focuses on making the complex and technical streamlined toward simplicity. We believe that these attributes make a Scrum-like model well suited to the process of identity development.

Furthermore, the Scrum model emphasizes both the agency and motivation required by individuals and groups and is all about giving people the proper environment and the support that they need to get the job done (i.e., the "job" here being building a computing identity). Much like individuals and groups develop products, we have found that Latina students develop their computing identities by going through a type of planning, implementation, review, and retrospect-type process. Their daily computing experiences are like sprints of a Scrum model in which they are testing out their identities, launching them, getting feedback, and then designing, redesigning and, in some cases, leaving the project (i.e., computing) entirely.

At the individual level, Latina undergraduate students engage in a three-step recursive process that emphasizes daily negotiations of identity and sense-making. This process consists of: (1) pre-college computing identity experiences, (2) college computing identity experiences, and (3) revised computing identity and outcomes.

*Pre-college Computing Identity Experiences*

At the far left, we have the pre-college computing identity experiences. These experiences are accumulated from birth through the end of high school. Prior to coming to campus, students may have engaged in daily identity negotiations and identity sense-making in order to continuously refine and redefine their computing identities. These pre-college identities may have been shaped by early computing, math, and logic experiences. In addition, experiences with family, friends, peers, teachers, and student organizations influence students' pre-college computing identity development.

From these experiences, Latina students may have anywhere between a well-formed computing identity to a non-existent one. Pre-college computing identity experiences can be described using the four key elements of role identity theory: interest, competence, performance, and recognition (Stryker & Burke, 2000). In this case, computing interest relates to a student's preference or affinity for computing content, tools, or subjects. Competence refers to the ways in which students understand the knowledge and skills of computing. Performance refers to how students are able to utilize their understandings and act out their computing identities. Recognition refers to a student's ability to recognize themselves and be recognized by others as a computer scientist.

*College Computing Identity Experiences*

In the middle, we have the college computing identity experiences, which are the heart of this conceptual model. While in college, Latina computing students continue to engage in daily identity negotiations and identity sense-making recursive experiences in which they continuously (re)define their computing identities. Daily identity negotiations refer to the activities and interactions that Latina students have regularly that influence their identity development. These negotiations can start the moment that a Latina student has interaction with college-level computing content or interactions.

College identities are shaped by computing experiences such as their introductory computing courses, homework assignments, and experiential learning environments (Peters, 2014; Peters, Berglund, Eckerdal, & Pears, 2014; Peters & Pears, 2012, 2013). Students' identity is also shaped by their interactions with individuals within computing such as faculty, peers, and industry partners, as well as outside of computing with family and community members. Identity sense-making refers to the ways in which Latina students evaluate those experiences and decide, consciously or subconsciously, whether they see themselves as future computer scientists. As undergraduates,

Latina students develop their computing identities around the four constructs of interest, competence, performance, and recognition. Within this context, interest might refer to a Latina student's draw and curiosity to learn more about technological advancements or significant computing topics. Competence might refer to a Latina student's ability to utilize computing tools or technical skills, such as programming. Performance could refer to a Latina student's ability to utilize that knowledge to complete a task or engage in acting the part of a computer scientist. Finally, recognition might refer to the ways in which she acknowledges herself as the type of person who does computing or the ways in which others see her as such.

*Revised Computing Identity Experiences & Outcomes*

At the far right, we have the revised computing identity experiences and outcomes. The term "revised" computing identity experiences refers to the idea that Latina students may have new or modified computing identities at this point in time. In addition, those computing identity experiences influence the eventual retention, persistence, and career outcomes of the student.

A student's computing identity may become more or less salient as time passes. A stronger salient identity means that students become more interested in computing during the course of their college years and see those interests have greater alignment with their own aspirations and abilities. Students feel as though they are competent in their grasp of the content and can fully utilize the knowledge that they have to complete computing tasks. In addition, students feel comfortable performing that identity and taking on the role of computer scientist. This allows them to recognize themselves as computer scientists as well as be recognized by others as such. Conversely, the student may feel a sense of disconnection from the role of a computer scientist in relation to one of the four aspects of identity. For instance, a student could feel disinterested in the role of a computer scientist due to the content, the degree, or failing to see how their interests connect to those of the field. Or they could struggle to feel competent with computing content, perhaps as a result of entering college without prior programming experiences because they did not have access to computer science courses in high school. If institutions fail to address inequities early and students struggle to gain competence in topics that they are assumed to have knowledge of coming in, disparities could persist. Students could also experience a lack of affirmation when they try to perform their role as a computer scientist in the presence of others; they might not be able to utilize the knowledge and tools to demonstrate their competence to others. Finally, Latina students might not see themselves as computer scientists or be recognized by others as such. The outcome of these identity experiences is twofold: identity strength and salience, and their willingness to continue in the role of computer scientist. A student's computing identity becomes stronger and more salient for those who can see themselves in this role; this sense of identity means that students either decide to stay in the field or not. As they make sense of their experiences and revise their computing identities to be stronger and more salient, they have the ability to recover quickly from challenges and, when challenged, have the ability to sustain their computing identities. In this way, computing identity is an integral element of resilience for Latina women in these fields, providing them a foundation and resolve to persist in their majors and on into the computing workforce.

## *Five Interweaving Systems Defining the Environment*

In our model of computing identity development for Latina students, we propose that all individual-level processes take place within the context of various systems. All computing identity experiences are shaped by five interweaving systems (shown as rectangles within the model). These interweaving systems are the greater environment in which first pre-college computing identity experiences happen; background and experiences shape these identity experiences; students enter college with a computing identity or not as many Latina students come to (i.e., decide to major in) computing post-college enrollment. These interweaving systems, taken from the Ecological Theory of Development (e.g. Bronfenbrenner, 2005; Tudge, Mokrova, Hatfield, & Karnik, 2009), include: individual, microsystem, mesosystem, exosystem, and macrosystem. Within our model, the individual level is where a student's innate sense of self lies and identity development takes place. The microsystem is a larger but still closely bound system made up of computing interactions such as those with faculty, advisors, and peers as well as community-based interactions with family, school, church, and health services. The mesosystem is the space in between the microsystems and the exosystem in which students make sense of multiple influences. For some students, this distinction between the microsystem (college/school) and the mesosystem (family/community/off-campus work) may be more pronounced because students spend more time on campus during college. The exosystem is made up of influences from the larger environments around them such as social services, neighbors, local politics, mass media, and industry. The macrosystem refers to the attitudes and ideologies of the greater culture(s) of which they are a part.

Each of these systems has the ability to influence computing identity processes at the individual level (as demonstrated with the bidirectional arrows reaching across all of the systems). These systems shape pre-college computing identity experiences of Latina students as well as college computing identity experiences, including the daily identity negotiations, sense-making, and revised computing identity and outcomes. For example, in our emerging research that examines the lived experiences of Latina students in computing (Rodriguez & Ramirez, 2020), we find that often a computing identity for Latina students is influenced by economic needs from various system levels. At the individual-, micro-, or meso-level, Latina students from low socioeconomic backgrounds may seek out these fields and build a sense of computing identity around their ideas that this path will financially benefit themselves as well as their families and communities. The exosystem and macrosystem levels may influence how Latina students may consider gender, racial, ethnic, class, or other identities as they shape their computing identities. In contrast, they may also take into consideration other factors, both positive and negative, such as national trends around diverse hiring in computing or patriarchal attitudes about the abilities of Latina women in computing. For instance, in emerging research (Rodriguez & Ramirez, 2020), we see that the college computing identity experiences of Latina women are influenced by larger intersectional oppressions related to both gender and race stemming from the existing exo- and macrosystems. The environments, attitudes, and ideologies of the greater cultures surrounding Latina students might at times minimize the capabilities of Latina women on the basis of their race and gender and attempt to marginalize them from computing fields. It is from these various system levels that Latina undergraduate students in computing can both draw upon community cultural wealth and funds of identity as assets and experience issues of intersectionality and multiple forms of oppression, as explained in the next two sections.

### Community Cultural Wealth & Funds of Identity

Too often, the Latinx communities that our students come from are seen as a hindrance rather than an asset. Many scholars have challenged this idea, and instead have suggested that Latinx communities can provide valuable forms of community cultural wealth and funds of identity (e.g. Esteban-Guitart & Moll, 2014; Yosso, 2005). We believe that these assets are extremely important within the context of computing identity development. Within our model, these two assets are designated with the left-side double-headed arrow that demonstrates bidirectional influence on the individual-level computing identity experiences, including daily identity negotiations and sense-making during college and influence of the student on the environments in which they live.

From the various system levels, the computing identity development of Latina undergraduate students is influenced by a Latina student's ability to leverage community cultural wealth (Yosso, 2005) and funds of identity (Esteban-Guitart & Moll, 2014). Emergent findings (Rodriguez & Ramirez, 2020) highlight the importance of leveraging familial and social capital (e.g., parents, friends, peers, teaching assistants, STEM identity-based organizations) to navigate and be successful in computing spaces in higher education. These assets enable Latina students in computing to bring their own various forms of knowledge, skills, abilities, and identities into the identity development process. Thus, we see students' families, home communities, and environments as being integral to the computing identity development process.

Community cultural wealth is an asset-based approach that refers to the range of knowledge, skills, and abilities that Students of Color bring to their education from their home communities (Yosso, 2005). Community cultural wealth made up of six types of capital: navigational, aspirational, linguistic, resistant, social, and familial. When applied to the computing identity development process, leveraging navigational capital might refer to a student's ability to make their way within computing spaces, particularly those which may be unsupportive to them and their communities (e.g., seeking tutoring, accessing financial aid, participating in clubs). In this context, aspirational capital might refer to a student's hopes for their computing future which drive them to persist despite facing educational inequities. For instance, many Latina students are driven by their aspirations and build their identities as computer scientists despite learning programming upon completing their first computing course (rather than in middle school or high school, like many of their peers). Linguistic capital can refer to the language and communication skills that students bring with them to the computing setting (e.g., bilingualism), whereas resistant capital might refer to the way students leverage their computing knowledge and experiences to create equitable outcomes for themselves and others, particularly for others in their group (e.g., being intentionally inclusive during group work). Social and familial capital may refer to how Latina students utilize the wisdom, values, and stories of their communities to access and navigate computer science while remaining engaged with their communities and ensuring that there is a path for others (e.g., reflecting on the struggles of their parents for strength or mentoring aspiring Latina computer scientists). These forms of capital are useful tools for Latina students to navigate this context and may provide tools for understanding students' aspirations and persistence in computing.

Coupled with community cultural wealth, Latina undergraduate students may also bring with them various funds of identity that shape their computing identities. Much like community cultural wealth, funds of identity shape the individual that a person becomes over time. Students bring with them social, cultural, institutional, geographical, and practical funds of identity that stem from their internalized funds of knowledge (Esteban-Guitart & Moll, 2014). For instance, Latina students may bring with them social or cultural funds of computing identity that shape the kind of computer scientist that they want to become (e.g., a Latina student becomes the kind of computer scientist that builds culturally relevant technological infrastructure for her community). Latina students might also bring with them institutional (e.g., elements of social institutions, like religion) or geographical funds of identity (e.g., physical and constructed features of their environments, like their hometown). For example, a student's geographical surroundings could inform the way that Latina computer science students think about computing infrastructure possibilities or challenges (e.g., rural vs. urban, mountain vs. coastal). In addition, Latina computing students may also bring practical funds of identity in to shape their computing identities, including knowledge gained from meaningful activities, interests, or hobbies (e.g., gaming knowledge or tinkering hobbies). Students' funds of identity directly influence their computing identity development and vice versa. For example, emerging research has shown (Rodriguez & Ramirez, 2020) that some Latina students possess practical funds of identity from their participation in a computing major as a Latina, which influences their desire to "give back" to their communities (a geographical fund of identity).

### Intersectionality (Multiple Forms of Oppressions)

While Latina students in computing have the ability to draw upon community cultural wealth and funds of identity, we also directly acknowledge in this model the presence of intersectionality and multiple forms of oppressions that may influence the computing identity development for these students. Latina undergraduate students can experience multiple forms of oppression stemming from their multiple marginalized identities (Carbado, Crenshaw, Mays, & Tomlinson, 2013; Cho, Crenshaw, & McCall, 2013; Crenshaw, 1991). As racially minoritized students, Latinas in computing may experience intersecting axes of privilege, domination, and oppression throughout their computing identity development process. Within our model, intersectionality and multiple oppressions are designated with the right-side double-headed arrow that demonstrates bidirectional influence on the individual-level computing identity experiences (similar to community cultural wealth and funds of identity) and influence of the student on the environments in which they live.

*Intersectionality*, a term coined by Crenshaw (1991), is defined as compounding oppressions that influence the ways in which individuals come to their environment. Women of color may experience a "double-bind" (Malcom, Hall, & Brown, 1976; Ong, Wright, Espinosa, & Orfield, 2011) in which they experience intersecting oppressions as a result of being both women and people of color in fields which have historically been dominated by White men. Our emergent findings (Rodriguez & Ramirez, 2020) demonstrate that Latina women in computing may experience such marginalization as a result of sexism (as women) and racism (as people of color) in the field.

In this vein, belonging to intersecting socially marginalized groups, Latina students in computing may experience difficulty in accessing resources or in finding role models and mentors, for example. Although this model examines a group of individuals with a shared gender and ethnic identity, we recognize that Latina students are diverse in regard to their nationality, citizenship, first-generation college student status, race, language, etc. As such, research aiming to understand the experiences of Latina students in computing necessitates a nuanced approach.

## Implications for Advancing Computing & Promoting Resilience

This chapter proposed a new, integrated conceptual framework for computing identity development for Latina undergraduate students. This framework drew upon several established theories in order to conceptualize how these college students experience identity development both at the individual level and within the environment around them. The framework acknowledges the influence of community cultural wealth and funds of identity that Latina students bring with them, as well as the forms of oppression that they may experience. It is our hope that this framework not only advances the field but also contributes to the ways in which we think about resilience for Latina students in computing fields. Developing a computing identity is an act of resilience and persistence despite challenges, as well as an act of resistance against various forms of structural oppression. In presenting this conceptual framework, we recognize the ongoing need for research, theory building, and praxis on computing identity development. First and foremost, we would like to reiterate that despite this model, we continue to acknowledge that we cannot treat Latinas as a monolithic group, as they are diverse (e.g., nationality, citizenship, first-gen status, race) and their experiences are nuanced and different. As such, we encourage others to take this model, use it, and refine it to capture the many nuances that exist among Latinas in computing. It is our hope that scholars, practitioners, policymakers, and other stakeholders will continue to refine our understanding of computing identity development in order to better serve Latina undergraduate students. In the future, scholars might consider longitudinal research studies that enable them to see identity development over time. Researching identity development over time and across critical junctures in the educational journey might further illuminate our understanding of computing identity development. In addition, scholars may see to conduct research studies and build theory across multiple educational contexts (e.g., Hispanic-serving Institutions, Historically Black Colleges and Universities, private elites) and with Latina undergraduate students who possess a variety of identities and backgrounds (e.g., generational status, geography, community college transfer).

Although this chapter centers on the creation of a theoretical model, there are many practical strategies that practitioners can use to promote resilience and encourage Latina students to develop their identities as computer scientists. For example, the model suggests that faculty and staff in computing departments recognize where their underrepresented student populations come from and recognize what students bring to computing (i.e., asset-based community cultural wealth and funds of identity) rather than what they do not bring (e.g., deficit-based pre-college programming experiences). From our emerging research (Rodriguez & Ramirez, 2020), we see that Latina students possess limited knowledge of computing fields and computing identity experiences; however, they draw on a wealth of knowledge from their communities and background experiences to shape their computing identities. Furthermore, they also develop their computing identities through involvement in computing organizations and benefit from attending computing events such as the Richard Tapia Celebration of Diversity in Computing (e.g., diversity from all backgrounds and ethnicities) or the Grace Hopper Celebration (e.g., gender diversity).

In turn, practitioners might look toward using real-life examples in computing classes that tackle social problems and challenges that people care about, as well as examples showing people from diverse backgrounds (e.g., Latinx population). In addition, practitioners might consider ways (e.g., assignments, first-year experience courses, culminating projects) in which they can encourage students to draw upon the rich experiences that they have had in the past in order to shape their computing identities. Practitioners might also consider ways in which they can support Latina students in connecting to computing organizations with diversity missions, including those mentioned earlier.

In contrast to these practical solutions for computing identity development, we would be remiss if we did not also challenge stakeholders to reconsider typical notions of resilience and their implications on broadening participation in computing. Without a commitment to the dismantling of sexism, racism, and other oppressions, our efforts to enhance computing identity and encourage resilience are futile. Rather than solely focusing on building computing for the purposes of resilience, we must also eliminate the structural oppressions that require resilience in the first place.

Future research and theory-building might also investigate how degree programs, curricula, and delivery influence computing identity. In the present, practitioners and policymakers can utilize what we already know about computing identity development for Latina undergraduate students to shape practice and policy. We can do this by thinking about computing experiences as identity creation experiences and using forms of community cultural wealth and funds of identity to actively engage students in this developmental process. We can also do this by building in mechanisms of identity development throughout the curriculum, internship, and career preparation experiences. Finally, better theory-building hinges on engaging scholars, practitioners, and policymakers in praxis, or the bringing together of research, theory, and practice for a better change process. Together, these groups might look to the conceptual framework presented here to take action on their campuses and engage in a reflective process in which they analyze the results, revise our understanding of computing identity development, and implement further changes to both our theoretical understanding of the process and its practical application.

## Conclusion

As the demand for computer technology grows in the next 50 years, the demand for computer scientists and a citizenry of people who are fluent and literate in computing will also continue to increase. The Latinx population is expanding at an unprecedented rate, and Latinas in particular have become a considerable market of higher-education consumers; as such, it is more important now than ever that Latinas are included in the movement to encourage resilience and train more computer scientists in computing.

This chapter addressed a review of the literature regarding Latinas in computing, as well as provided a theoretical framework that helps better understand how Latina computing identity is developed and nurtured through multiple layers, which include: (1) iterative computing identity development at the individual level; (2) five interweaving systems that define the computing and broader environment; (3) community cultural wealth and funds of identity derived from those environments; and (4) a distinct recognition of intersectionality and multiple forms of oppression these women may experience. This framework is intended to provide both insight and critique into how Latinas' computing identities interact with, and are ultimately shaped by, the various systems in computing technology.

## References

Barker, L. J., McDowell, C., & Kalahar, K. (2009, March). Exploring factors that influence computer science introductory course students to persist in the major. In *ACM SIGCSE Bulletin* (Vol. 41, No. 1, pp. 153–157). ACM.

Bronfenbrenner, U. (2005). *Making human beings human: Bioecological perspectives on human development*. Thousand Oaks, CA: Sage.

Carbado, D. W., Crenshaw, K. W., Mays, V. M., & Tomlinson, B. (2013). Intersectionality: Mapping the movements of a theory. *Du Bois Review: Social Science Research on Race, 10*(2), 303–312.

Cheryan, S., Plaut, V. C., Davies, P. G., & Steele, C. M. (2009). Ambient belonging: How stereotypical cues impact gender participation in computer science. *Journal of Personality and Social Psychology, 97*(6), 1045–1060.

Cho, S., Crenshaw, K. W., & McCall, L. (2013). Toward a field of intersectionality studies: Theory, applications, and praxis. *Signs: Journal of Women in Culture and Society, 38*(4), 785–810.

Crenshaw, K. W. (1991). Mapping the margins: Intersectionality, identity politics and violence against women of color. *Stanford Law Review, 43*(6), 1241–1299.

Esteban-Guitart, M., & Moll, L. C. (2014). Funds of identity: A new concept based on the funds of knowledge approach. *Culture & Psychology, 20*(1), 31–48.

Lewis, K. L., Stout, J. G., Pollock, S. J., Finkelstein, N. D., & Ito, T. A. (2016). Fitting in or opting out: A review of key social-psychological factors influencing a sense of belonging for women in physics. *Physical Review Physics Education Research, 12*(2), 1–10.

Lynch, M. (2016). *The tech divide: An opportunity gap schools must close*. Retrieved from www.theedadvocate.org/the-tech-divide-an-opportunity-gap-schools-must-close/

Malcom, S., Hall, P., & Brown, J. (1976). *The double bind: The price of being a minority woman in science*. Washington, DC: American Association for the Advancement of Science.

Margolis, J., Estrella, R., Goode, J., Holme, J. J., & Nao, K. (2008). *Stuck in the shallow end: Education, race, and computing.* Cambridge, MA: MIT Press.

National Center for Education Statistics (NCES) (2017). *Digest of education statistics: Bachelor's degrees conferred by postsecondary institutions, by race/ethnicity and sex of student: Selected years, 1976–77 through 2015–2016.* Retrieved from https://nces.ed.gov/programs/digest/d17/tables/dt17_322.20.asp?current=yes

National Science Foundation (2016). National center for science and engineering statistics, special tabulations of U.S. Department of education, national center for education statistics, integrated postsecondary education data system, completions survey. Bachelor's degrees awarded to women by field, citizenship, ethnicity, and race. Table 5–4.

National Science Foundation (2017). *Integrated Postsecondary Education Data System (IPEDS) Completion Survey by Race, Integrated Science and Engineering Resources Data System (WebCASPAR).* Retrieved from https://ncsesdata.nsf.gov/webcaspar/

National Science Foundation (2019). *Women, minorities, and persons with disabilities in science and engineering.* Retrieved from www.nsf.gov/statistics/2017/nsf17310/digest/fod-wmreg/hispanic-women-by-field.cfm

Ong, M., Wright, C., Espinosa, L. L., & Orfield, G. (2011). Inside the double bind: A synthesis of empirical research on undergraduate and graduate women of color in science, technology, engineering, and mathematics. *Harvard Educational Review, 81*(2), 172–208.

Peters, A. K. (2014). *The role of students' identity development in higher education in computing* (Doctoral dissertation). Uppsala Universitet, Uppsala.

Peters, A. K., Berglund, A., Eckerdal, A., & Pears, A. (2014). First year computer science and IT students' experience of participation in the discipline. In *2014 international conference on teaching and learning in computing and engineering* (LaTiCE) (pp. 1–8). Kuching: IEEE.

Peters, A. K., & Pears, A. (2012). Students' experiences and attitudes towards learning computer science. In *2012 Frontiers in education conference* (FIE) (pp. 1–6). Seattle, WA: IEEE.

Peters, A. K., & Pears, A. (2013). Engagement in computer science and IT—What! A matter of identity? In *2013 International conference on teaching and learning in computing and engineering* (LaTiCE) (pp. 114–121). IEEE.

Pew Research Center. (2017). *Four-in-ten Americans credit technology with improving life most in the past 50 years.* Retrieved from www.pewresearch.org/fact-tank/2017/10/12/four-in-ten-americans-credit-technology-with-improving-life-most-in-the-past-50-years/

Rodriguez, S. L., & Lehman, K. (2018). Developing the next generation of diverse computer scientists: The need for enhanced, intersectional computing identity theory. *Computer Science Education, 27*(3–4), 229–247.

Rodriguez, S. L., & Ramirez, D. (2020, January). *Becoming a computer scientist: Enhancing education for Latina students in computing.* Presentation at the Joint National Conferences of NAAAS, NAHLS (National Association of Hispanic & Latino Studies), NANAS, & IAAS, Dallas, TX.

Sax, L. J., Blaney, J. M., Lehman, K. J., Rodriguez, S. L., George, K. L., & Zavala, C. (2018). Sense of belonging in computing: The role of introductory courses for women and underrepresented minority students. *Social Sciences, 7*(8), 122–145.

Sax, L. J., Zimmerman, H., Blaney, J. M., Toven-Lindsey, B., & Lehman, K. (2017). Diversifying undergraduate computer science: The role of department chairs in promoting gender and racial diversity. *Journal of Women and Minorities in Science and Engineering, 23*(2), 101–119.

Stout, J., & Blaney, J. M. (2018). "But it doesn't come naturally": How effort expenditure shapes the role of growth mindset on women's intellectual belonging in computing. *Computer Science Education, 27*(3–4), 215–228.

Stryker, S., & Burke, P. J. (2000). The past, present, and future of an identity theory. *Social Psychology Quarterly, 63*(4), 284–297.

Team Linchpin (2019). *A beginner's guide to the agile method & scrums.* Retrieved from https://linchpinseo.com/the-agile-method/

Tudge, J. R., Mokrova, I., Hatfield, B. E., & Karnik, R. B. (2009). Uses and misuses of Bronfenbrenner's bioecological theory of human development. *Journal of Family Theory & Review, 1*(4), 198–210.

Varma, R. (2010). Why so few women enroll in computing? Gender and ethnic differences in students' perception. *Computer Science Education, 20*(4), 301–316.

Yosso, T. J. (2005). Whose culture has capital? A critical race theory discussion of community cultural wealth. *Race Ethnicity and Education, 8*(1), 69–91.

# 3 The Pathway to the PhD

## Latinas as STEM Doctorates From 1975–2010

*Frank Fernandez, Hyun Kyoung Ro,
Miranda Wilson, and Veronica Crawford*

Doctoral programs help students develop competence and expertise so they can become leaders in a variety of professions (Parsons & Platt, 1968; Woodrow Wilson National Fellowship Foundation, 2005). In the twenty-first century, it is especially important to equitably develop doctoral-level talent in STEM fields. Leading up to US involvement in World War II, STEM fields began to receive "prominent attention" over humanities and social sciences because science and technology fields were seen as the "handmaidens of economic interests" (Parsons, 1946, p. 458). Within a few decades after World War II, US STEM doctoral education expanded almost exponentially, which allowed the United States to lead the world in producing STEM research (Fernandez & Baker, 2017; Powell, Baker, & Fernandez, 2017). Universities that had previously focused on undergraduate education began offering STEM PhDs, and old providers expanded the size of their programs. Despite unprecedented growth in the capacity for STEM training, the doctoral education system never equitably included US-born people of color (Fernandez, Baker, Fu, Muñoz, & Ford, 2017).

The United States *should* increase STEM doctoral educational opportunities for Latinas as a matter of social equity and fairness. The United States *must* increase STEM doctoral opportunities for Latinas to meet labor market demands (Malcom, Dowd, & Yu, 2010). The US Bureau of Labor Statistics projected that the number of STEM jobs will increase by 28.2% between 2014 and 2024—more than quadruple the growth in other occupations (Fayer, Lacey, & Watson, 2017). As STEM occupations continue to expand, the percentage of the US population that identifies as White alone continues to decline; in fact, the majority–minority demographic crossover is projected to occur by 2044 (Colby & Ortman, 2015). Based on national projections, it is crucial that underrepresented populations, like Latinas, be brought into the STEM labor market (Martínez & Gayfield, 2019).

In addition to meeting the needs of the twenty-first century labor market, training more Latina STEM PhDs will help expand the pool of Latinas who can work in college and university settings as role models and mentors. Latinas are underrepresented among college and university faculty (Turner,

González, & Wood, 2008). Increasing the number of Latina faculty is critical given that Students of Color are more likely to complete courses and earn better grades when they take courses with faculty of color (Fairlie, Hoffman, & Oreopoulos, 2014). Increasing Latina doctorates who take faculty positions can help improve student success for the growing ranks of Latina and Latino undergraduate students. The role of Latina faculty is especially important in STEM undergraduate education, because women of color in STEM fields often lack role models and experience microaggressions that prevent them from persisting in STEM fields (Espinosa, 2011). Therefore, for both economic and educational purposes, it will advance the national interest if the US higher education system draws on the talents of all its people, including by recruiting, retaining, and improving doctoral attainment rates of Latinas.

In this chapter, we examine the pathway for Latinas who earned STEM PhDs between 1975 and 2010. We examine the representation of Latinas who earned STEM PhDs, relative to their share of the population. Additionally, we consider whether there was a change in the percentage of Latina STEM undergraduates who went on to earn STEM PhDs. We inform this descriptive study by reviewing prior literature. After presenting our findings, we provide recommendations for increasing opportunities for Latinas to earn STEM PhDs—a key credential for moving into positions of leadership throughout our society.

## Prior Literature on Latinas in Graduate Education

Prior studies established that Latinas were underrepresented among doctorate degree earners, relative to their share of the population. Data from the 1980s demonstrated that Latinas (i.e., Chicanas or Latinas of Mexican descent) were less well represented than other racial groups across multiple fields, including engineering, life sciences, and hard sciences (Solórzano, 1994), as well as social sciences (Solórzano, 1995).[1] The same disparities persisted through the 1990s (Watford, Rivas, Burciaga, & Solórzano, 2006). Using data from the national Survey of Earned Doctorates, Watford et al. (2006) examined doctoral attainment of students with respect to gender within each racial group. They found an increase in Latina doctoral recipients from 3.5% to 5% during the 11-year period (1990–2000); however, the gains were relatively small in relation to the overall population growth of Latinas in the United States over the same period. Additionally, Latinas continued to face both overt and covert forms of marginality in pursuit of doctoral education (Watford et al., 2006). Recent work has examined Latina and Latino doctorate production in the social sciences during the 2000s (Fernandez, 2018; Fernandez, 2019; Fernandez, 2020), but additional research is needed that examines the long-term pathway between Latina STEM undergraduates and PhD earners.

If Latinas and Latinos are to participate in the social and economic growth of this society, institutions of higher education must seek ways to increase educational attainment rates for this group (Contreras & Gándara, 2006). In the remainder of this section, we review literature on the pathway to doctoral education. Our review of prior scholarship both informs our findings and challenges institutions to consider ways to train more Latina STEM PhDs.

### A History of Prejudice Against Latinas in the US Educational System

If Latinas and Latinos are underserved or excluded in the early stages of the educational system, then it is not surprising if they are poorly represented at the opposite end of the system. Latinas and Latinos have historically been underrepresented throughout the educational pipeline, both relative to their share of the population and compared to other ethnic and racial groups. Latina and Latino underrepresentation is not an accident; Latinas and Latinos were historically excluded and marginalized by the American educational system. Before the famous *Brown v. Board of Education* (1954) desegregation case, the Lemon Grove School District built a segregated two-room schoolhouse in the suburbs of San Diego, California for Latina and Latino children. In January 1931, the principal of Lemon Grove Grammar School physically blocked Latina and Latino students from entering their former school and directed them to the new segregated building. At the state level, legislators introduced a bill to support segregating Latina and Latino students in Lemon Grove (Alvarez, 1986).

The families of the Latina and Latino students sued. When the Superior Court of California in San Diego indicted the board members of the Lemon Grove School District, the board members claimed that they were not trying to maliciously segregate Latina and Latino students. Instead, the trustees justified their actions by claiming that segregation would improve education for Latina and Latino children. By March 1931, the judge ruled that the Lemon Grove School District had violated state laws and ordered desegregation of the grammar school (Alvarez, 1986).

Like the trustees in Southern California, South Texans also defended treating Latina and Latino schoolchildren differently than their White peers. A few years after *Brown*, the League of United Latin American Citizens sued the Driscoll Consolidated Independent School District for the way it treated Latina and Latino students. The South Texas district made it a practice to hold Latina and Latino children back from advancing beyond the first grade; some children were kept in first grade for up to four years. Although the Driscoll district denied that it had outrightly segregated Latinas and Latinos (all of whom were classified as White under the law), the court found that the district had denied Latina and Latino students equal protection (Godfrey, 2008).

*Further Challenges in the Pathway to the PhD*

The challenges in primary and secondary schools continue to the undergraduate level. Poor high school preparation leaves many Latinas underprepared for rigorous STEM courses, and makes it difficult for them to succeed so they can pursue doctoral aspirations (Ruiz, 2013). Additionally, a quantitative multi-campus study shows that many colleges and universities, particularly highly selective institutions, have inhospitable climates for women of color. The negative climate of selective institutions is at least partly a function of a lack of ethnic diversity in STEM classrooms, laboratories, and departments (Espinosa, 2011). For example, Latina undergraduates in STEM departments too often lack role models because there are very few Latina graduate students or professors (Espinosa, 2011).

There are several challenges at the undergraduate level that can discourage Latinas from pursuing their post-baccalaureate plans. For example, a mixed-method study of Latinas and Latinos in engineering programs found that they tended to have higher aspirations to attend graduate school after earning bachelor's degrees compared to other racial or ethnic groups. Yet, the Latina and Latino engineering undergraduates reported that they were less exposed to information about graduate programs (Gonzalez, 2015). Additionally, students who take on loans to finance baccalaureate studies have lower odds of attending graduate programs. Among Latinas and Latinos, the likelihood of enrolling in post-baccalaureate programs declines as debt levels increase (Malcom & Dowd, 2012).

Scholars have analyzed autobiographical essays and found that Latina master's and doctoral students experience similar challenges as Latina STEM undergraduates (Rendón, Nora, Bledsoe, & Kanagala, 2019). In addition to culture shock and feelings of tokenism—brought on by being the only Latina, or sometimes the only woman, in the classroom—Latinas experience academic-related challenges throughout their paths to the PhD. Academic challenges may include a lack of information about how to succeed in STEM, pursuing graduate study if English is a second language, difficulty with scholarly writing, and lack of information about presenting and publishing for academic audiences. To complicate all that, Latinas often face life-stage challenges as they move from undergraduate to doctoral programs. For example, life challenges can include the onset of health problems, caring for siblings or other family dependents, and financial challenges—recall that many federal grants are reserved for undergraduate programs and federal aid is limited for graduate study (Rendón et al., 2019).

Universities continue to struggle with how to restructure graduate programs so they offer equitable opportunities for racial minority students. For instance, graduate diversity officers have identified challenges in improving the admissions process, which is closely tied to faculty support and offering funding to doctoral students (Griffin, Muñiz, & Espinosa, 2012; Posselt, 2016). After admission, racial minority students can often struggle in doctoral programs. One important study found that minority students were the targets of microaggressions and experienced "vicarious trauma" when they heard about racist statements that were directed at others (Slay, Reyes, & Posselt, 2019, p. 272). Microaggressions and vicarious traumas created challenges to minority students' success in their graduate programs (Slay et al., 2019). In environments that lack diversity and that can be overtly hostile, doctoral students need academic support, psychosocial support, and sociocultural support from their advisors (Posselt, 2018).

## New Findings on Latinas as STEM PhD Earners

### Data

We analyzed national data to explore STEM doctoral pathways among Latinas. First, we used the American Community Survey (ACS) to examine women's educational pathways from elementary school enrollment to doctoral degrees by five racial groups (Latinas, African Americans, Native Americans, Asian Americans, and Whites). The ACS includes census data collected annually and information about educational attainment in the United States. The ACS is a vital tool used by community stakeholders to gather information about population dynamics in communities (U.S. Census Bureau, 2019).

Second, to present the number of Latinas who earned degrees by STEM disciplines at the associate's, bachelor's, and master's levels, we analyzed data from the National Center for Science and Engineering Statistics (NCSES, 2011, 2019a, 2019b). These data include information from a compilation of NCSES surveys that are used to report the characteristics of women, minorities, and persons with disabilities in science and engineering. Finally, we analyze data from the Survey of Earned Doctorates (SED). We limit our analysis of SED data to Latinas who were US citizens or permanent residents. For our analysis of SED, we define STEM fields as including computer information sciences, engineering, life sciences, math, and physical sciences (NORC, 2012).

### Educational Pathways

Informed by prior literature, we begin by updating a previous snapshot on the Latina educational pipeline to the PhD—for all academic fields. Figure 3.1 demonstrates that Latinas navigate an inequitable educational system. Many high schools fail to graduate Latina students at similar levels as other groups—especially White women—and the pathway to the PhD continues to narrow as Latinas enter and move through the higher education system. In 2000, Latinas had the lowest high school graduation rates compared to African Americans, Native Americans, Asian Americans, and Whites as a legacy of public schools marginalizing Latina and Latino students (Pérez

## The US Female Educational Pipeline by Race: 2006-2010

| Latinas | African Americans | Native Americans | Whites | Asian Americans |
|---|---|---|---|---|
| 100 Elementary School Students | 100 Elementary School Students | 100 Elementary School Students | 100 Elementary School Students | 100 Elementary School Students |
| 63 Graduate From High School | 82 Graduate From High School | 80 Graduate From High School | 90 Graduate From High School | 84 Graduate From High School |
| 14 Graduate From College | 19 Graduate From College | 14 Graduate From College | 30 Graduate From College | 48 Graduate From College |
| 4 Graduate From Graduate School | 7 Graduate From Graduate School | 5 Graduate From Graduate School | 11 Graduate From Graduate School | 17 Graduate From Graduate School |
| 0.4 Graduate With Doctorate | 0.5 Graduate With Doctorate | 0.3 Graduate With Doctorate | 0.9 Graduate With Doctorate | 2.0 Graduate With Doctorate |

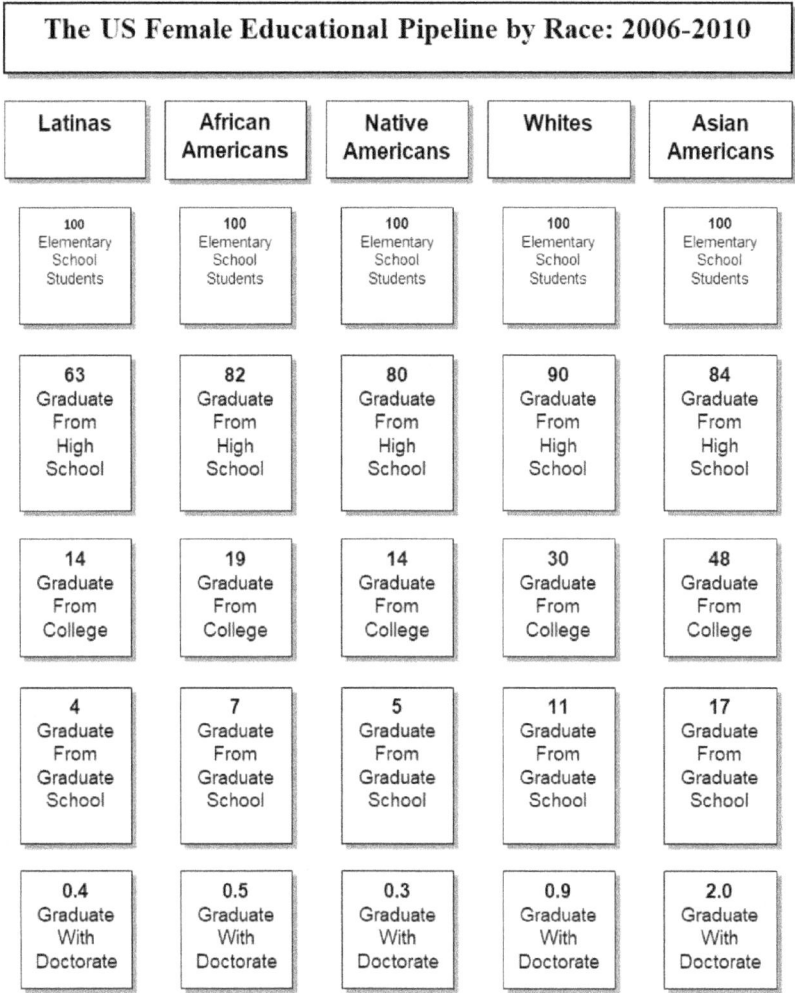

*Figure 3.1* Latinas' Attainment Rates at Each Stage of the US Female Pathway to the PhD.

Source: US Bureau of the Census, 2006–2010 population estimates from the American Community Survey. The five-year estimates are the best source of national demographics between the decennial census.

Note: Figure organized in the format previously used by Pérez Huber et al. (2006) and Watford et al. (2006).

Huber, Huidor, Malagón, Sánchez, & Solórzano, 2006). One decade later, Latinas were still the least likely racial group to graduate from high school. For every 100 students who entered the US educational system, fewer Latinas graduated from college than African American, Asian American, and

*Table 3.1* Number of Latinas Who Earned STEM Degrees by Field and Degree Level.

| STEM Field | Associate's | Bachelor's | Master's |
|---|---|---|---|
| Agricultural Sciences | 36 | 573 | 122 |
| Biological Sciences | 357 | 4,356 | 319 |
| Computer Sciences | 759 | 590 | 159 |
| Earth, Atmospheric, & Ocean Sciences | 3 | 127 | 33 |
| Mathematics and Statistics | 81 | 411 | 56 |
| Physical Sciences | 145 | 563 | 65 |
| Astronomy | N/A | 15 | 2 |
| Physics | N/A | 49 | 12 |
| Chemistry | N/A | 484 | 47 |
| Engineering | 55 | 1,346 | 366 |

Note: This table is not inclusive of all STEM fields and sub-fields within NCSES. The NSF encompasses a broader definition of STEM including all fields and sub-fields within mathematics, engineering, computer and information sciences, natural sciences, and social/behavioral sciences; under the broader STEM definition, sub-fields included areas like economics, sociology, psychology, political science, architecture, communication, and family and consumer sciences/human sciences (Congressional Research Service, 2012; National Science Board, 2018).

White women. The inequities throughout the educational system culminated in Latinas being underrepresented at the end of the pathway to the PhD, where smaller percentages of Latinas earned doctorates than African American, Asian American, and White women. In 2000, similar percentages (0.3%) of Latinas and African American women earned doctorates (Pérez Huber et al., 2006). Although both groups of women made progress during the decade, by 2010 Latinas fell behind African American women.

Table 3.1 reports the absolute number of Latinas who earned degrees by STEM category at the associate's, bachelor's, and master's levels based on data from the NCSES (2011, 2019a). In 2010, of all 25,743 science and engineering (S&E) associate's degrees awarded to women, 4,031, or 15.7%, went to Latinas. Of all 264,283 S&E bachelor's degrees awarded to women, 26,000, or just under 9.9%, were earned by Latinas (NCSES, 2019a); and of all 63,660 S&E master's degrees awarded to women, 4,026—approximately 6.3%—were earned by Latinas (NCSES, 2019b). Note that at all levels below the doctorate, Latina STEM degree attainment varies widely across fields.

### Latina STEM Doctorate Production

Figure 3.2 plots annual changes in Latina PhD production between STEM and non-STEM fields. In 1975, around 5% of the doctorates Latinas earned were in STEM fields. In Figure 3.2, we focus on Latinas as PhD earners. Even if the *number* of Latina PhD earners has increased, the *percentage* of PhDs that were earned by Latinas in all fields did not change much over 35 years. Between 1975 and 2010, the number of PhDs awarded to US-born Latinas increased from

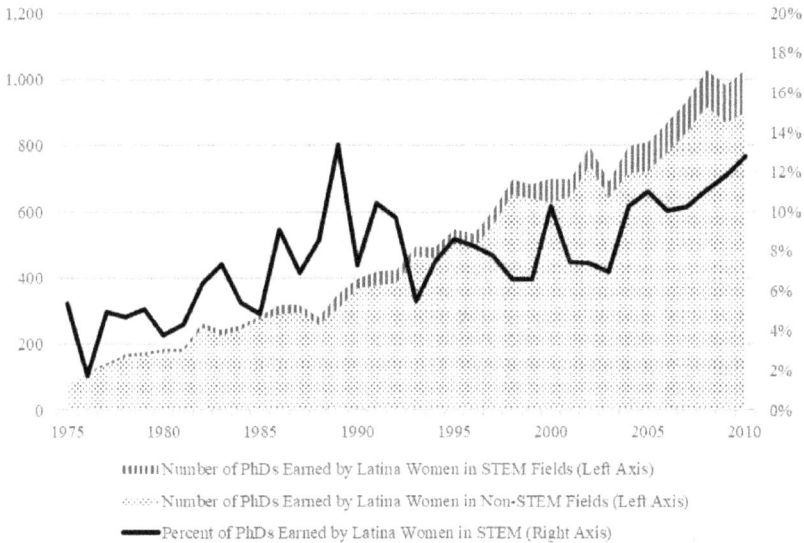

ıııııNumber of PhDs Earned by Latina Women in STEM Fields (Left Axis)

········Number of PhDs Earned by Latina Women in Non-STEM Fields (Left Axis)

▬▬▬Percent of PhDs Earned by Latina Women in STEM (Right Axis)

*Figure 3.2* Latina PhD Production, 1975–2010.

Note: Figure 3.2 is based on authors' calculations of restricted data from the Survey of Earned Doctorates (SED). For our analysis of SED, we defined STEM fields as including computer information sciences, engineering, life sciences, math, and physical sciences (NORC, 2012).

a little more than 70 to more than 1,000. Figure 3.2 plots the annual number of US-born Latinas who earned PhDs along the left axis.

By 2010, approximately 13% of Latina PhDs graduated from STEM programs. However, the 2010 percentage was still just below the 1989 zenith. One could argue that the percentages of PhDs earned in STEM stayed relatively low because production of Latina PhDs increased so quickly in non-STEM fields. However, analyzing the same data source, Fernandez (2018) revealed that Latinas are still relatively underrepresented relative to the population in social sciences and humanities.

Other work suggests that the relative drop of US-born Latinas earning STEM PhDs after 1989 coincides with a period when American universities significantly increased enrollments of international students in STEM PhD programs (Fernandez et al., 2017). Of course, we do not mean to imply that international PhD enrollments squeezed Latinas out of STEM programs. Yet, it is evident that even though the US system of STEM doctoral education continued to expand, and the percentage of Latinas in the population continued to increase, US universities did not make sufficient progress in expanding Latina STEM PhD production.

In summary, Figure 3.2 shows that Latinas earned greater numbers of PhDs in both STEM and non-STEM fields between 1975 and 2010.

However, the relative percentage of PhDs earned by Latinas in STEM fields dipped several times since the late 1980s. Two full decades passed before Latinas again earned a similar percentage of STEM PhDs.

Next, we consider the pathway between undergraduate STEM programs and STEM PhD programs. For Figure 3.3, we examine Latina PhD earners in the Survey of Earned Doctorates for whom undergraduate degree information was also available. Latinas who earned undergraduate degrees in STEM but did not earn PhDs are not included in Figure 3.3. In only five years during the 35-year period between 1975 and 2010, did at least half of Latinas who earned doctorates graduate from both STEM undergraduate and doctoral programs; those years were outliers in which the number of Latina PhDs who earned STEM undergraduate degrees never exceeded single digits. In most years, Latina PhDs who earned STEM undergraduate degrees completed their doctoral studies in non-STEM fields (see Figure 3.3).

On one hand, we should celebrate every Latina who earns a PhD in any field. There are certainly cases of people who earn undergraduate degrees in STEM and then apply their STEM knowledge and skills in other fields (Ro, Lattuca, & Alcott, 2017). For example, a Latina might find undergraduate training in biology to be useful in a PhD program in business. On the other hand, it is problematic for STEM graduate education and industries if Latina STEM undergraduates desire to leave the STEM fields because of negative

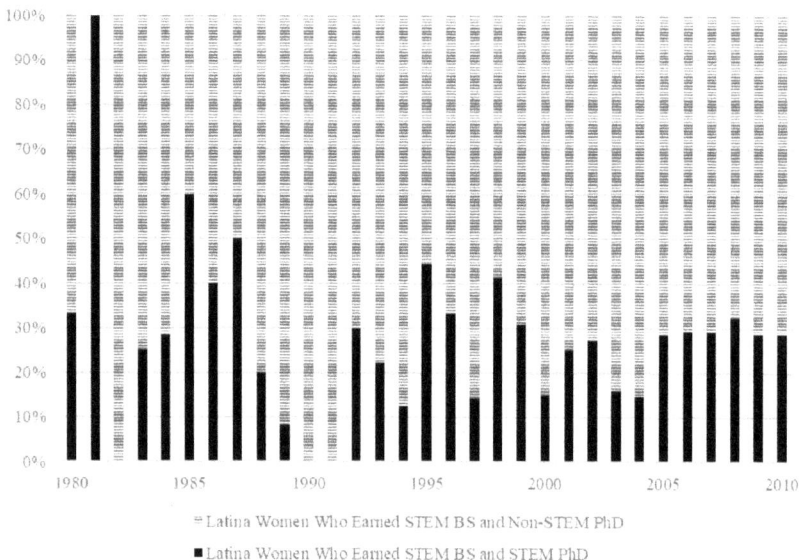

*Figure 3.3* Lack of Retention in STEM Among PhD Earners, 1980–2010.

Note: Figure 3.3 is based on authors' calculations of restricted data from the Survey of Earned Doctorates.

experiences or chilly climates (Espinosa, 2011). Anecdotally, people often assume that there are few Latina STEM faculty members or few Latinas in STEM PhD programs because there are too few Latinas in undergraduate programs—or too few going to graduate school. Yet the data in Figure 3.3 show in many years, Latinas who were trained in STEM doctoral programs demonstrated their ability to complete graduate coursework, and to write and then defend dissertations.

The descriptive data in Figure 3.3 do not give us any indication of why Latinas with STEM undergraduate degrees chose to earn PhDs in non-STEM fields. While it may be that they experienced chilly or hostile climates in their undergraduate studies and thus chose to pursue different fields for doctoral studies, they may have chosen non-STEM disciplines but still incorporated their STEM knowledge and skills into their graduate education (Ro et al., 2017). It may also be that they chose graduate schools or programs based on life-stage factors, such as being location-bound for work or family reasons (Grimes & Morris, 1997; Warnock & Appel, 2012). The data in Figure 3.3 lead to a need for further research on the pathways that Latinas take to PhDs and their decisions to continue to study STEM fields or to move into non-STEM subjects.

## Limitations

We used descriptive data to identify national trends in Latina STEM degree attainment between 1975 and 2010. We also provided snapshots of the educational pipeline (as seen in Figure 3.1 and Table 3.1) toward the end of the first decade of the twenty-first century. Prior studies examined shorter periods of time or provided snapshots with older data (e.g., Solórzano, 1994). The longer trends in this chapter demonstrate numerical but not proportionate increases of Latina STEM PhD earners, and the more recent snapshots update seminal work (e.g., Pérez Huber et al., 2006; Solórzano, 1994; Watford et al., 2006). However, the data in this chapter do not include the most recent decade of Latina PhD production in STEM fields. Along with the other chapters in this volume, we hope that this chapter provides a foundation for scholars and policymakers to gather and analyze more recent data on Latina PhD earners.

## Recommendations for Improving Pathways to STEM Doctorates

During the previous decade, the National Institute of General Medical Sciences and the Howard Hughes Medical Institute convened a Joint Working Group on Improving Underrepresented Minorities Persistence in Science, Technology, Engineering, and Mathematics (Estrada et al., 2016). While that group did not focus on Latinas, they asserted that colleges and universities should place greater emphasis on removing institutional barriers and developing interventions that "lift" students' participation and persistence in STEM fields. For example, the committee found evidence that several strategies increase persistence at the undergraduate and graduate levels. They found that STEM departments should seek to create strategic partnerships between program directors and university stakeholders to create and implement interventions. For example, partners can work to support students' academic preparation or interest and motivation in pursuing STEM degrees and careers. STEM departments and programs should also work to implement active learning pedagogies within STEM curriculums, address student resource disparities, and help students see STEM subjects as inherently creative and meaningful disciplines in ways that align with students' dispositions. Furthermore, Estrada et al. (2016) suggested that institutions should engage in action research methodologies to track successes and failures at the institutional level while collecting data to provide evidence-based interventions that aid in reducing disparities.

Other scholars echo and expand upon the recommendations in the joint working group report. Rendón and colleagues (2019) emphasize strategies that focus on early stages in the pathway to the doctorate. They challenge colleges and universities to increase access to undergraduate programs by supporting early preparation programs (e.g., first-year experience courses for STEM majors), remarketing their recruitment materials to Latina and Latino communities, and avoiding making admissions or advising decisions based on racialized, deficit-based views about Latina and Latino students and communities. Additionally, Rendón et al. (2019) encourage faculty to promote Latina and Latino student success by redesigning curricula and pedagogy, providing professional development for faculty and staff to help recruit and retain Latina and Latino students, and helping students develop both academic and personal competencies.

Increasing positive learning experiences and learning outcomes could be key in promoting Latinas to pursue doctoral education in STEM. In terms of pedagogy and competencies, prior work suggests that Latinas in engineering programs may excel at developing leadership skills as undergraduates (Ro & Loya, 2015), which may help them pursue advanced degrees and ultimately become leaders in STEM fields. Faculty can continue to improve pedagogical practices in STEM by creating learning environments that promote peer interactions, co-curricular involvement, and access to undergraduate research opportunities (Ro & Knight, 2016; Ro, Knight, & Loya, 2016). Faculty can also support cross-disciplinary research and training opportunities between STEM departments and gender or ethnic studies departments. Strategic cross-disciplinary partnerships can help Latina students develop complementary competencies, such as an understanding of social structures and the influence of social institutions on women's intellectual development and access to scientific material (Espinosa, 2011). Espinosa (2011) argues that partnerships between STEM and gender or ethnic studies faculty are warranted and represent a lost opportunity if institutional leaders fail to forge and promote such connections.

Provosts and deans should work to hire Latina/o faculty and administrators, who play a vital role in the academic achievement of Latina students. A study conducted by Torres and Hernandez (2009) found that undergraduate Latina and Latino "students with an advisor/mentor consistently have higher levels of institutional commitment, satisfaction with faculty, academic integration, cultural affinity, and encouragement" (p. 1). Considering the effects of role models in Latina/o education, it is important to also note the gender inequalities that exist in STEM fields. Garcia (2006) revealed the unique challenges that Latinas face while pursuing doctorates in male-dominated STEM fields. The biggest issue for Latina students is finding faculty mentors who share similar experiences and can offer guidance and support to meet their educational needs (Garcia, 2006; Ruiz, 2013).

### Recommendations for Graduate Schools and Graduate Program Faculty

Graduate program coordinators should focus on helping to retain racial minority students. First, deans and diversity officers should strive to improve the campus climate (i.e., whether racial minority students feel welcome or ostracized). To improve the climate of graduate programs, faculty and staff who work with graduate students can be sensitive to the psychological needs of underrepresented populations so that they can support and promote student success—which, in turn, will support future recruitment (Griffin et al., 2012).

Faculty can adopt specific advising practices to improve climate and support doctoral students' success. For instance, doctoral faculty can provide greater hands-on supervision to doctoral students and normalize struggles while cultivating a growth mindset. Often, tenured faculty turn over slowly, and departments should encourage faculty to develop cultural competencies and interpersonal skills for mentoring younger and more diverse cohorts of doctoral program admits. Finally, graduate schools should rethink cultural norms and practices, particularly those that factor into the admissions process and how programs encourage doctoral students to pursue faculty careers (Posselt, 2018).

University chief diversity officers and institutional researchers should investigate departments that have a history of excluding Students of Color during admissions and funding decisions. Conversely, university leaders can work to gain a contextual understanding of university-specific factors that promote full inclusion of students across gender, race, age, and other social identities (Slay et al., 2019). Of the factors identified by Slay and colleagues (2019), offering financial support to Latina and Latino doctoral students may be especially important. Millett and Nettles (2006) and Fernandez (2019) highlight the importance of providing teaching and research assistantships to Latina and Latino students.

### Policy Considerations

In the policy realm, federal policymakers can work to undo the damage that ballooning student debt has done in discouraging racial minority students from pursuing graduate degrees (Malcom & Dowd, 2012). Congress can support the pathway to graduate and professional education by reducing an over-reliance on loans. Financial aid policy reforms could include strengthening income-based repayment programs, reducing the debt burden ceiling, and increasing means-tested grant aid. Additionally, the federal government could increase graduate fellowship opportunities and other sources of non-loan support for Latinas and other minority students (Malcom & Dowd, 2012).

In addition to helping individual students, Congress and state policymakers may direct new and existing resources toward institutional programs that directly serve women in STEM (Espinosa, 2011). More specifically, Griffin and colleagues (2012) point out that organizations like the National Science Foundation and the National Institutes of Health should expand funding opportunities aimed at helping STEM programs recruit and retain Latinas and other Students of Color. At the state level, legislators and academic leaders can work to balance or counter anti-affirmative action policies by providing financial incentives to public institutions that enroll diverse graduate student populations and produce doctorates from underrepresented backgrounds (Griffin et al., 2012).

### Conclusion

The PhD remains an important milestone in the pathway to positions of leadership in academia, government, and industry. Fortunately, over the last few decades, the US system of graduate education has produced larger *numbers* of Latina doctorates in STEM. Unfortunately, in relative terms, the medium- to long-term trends fail to demonstrate significant advances in STEM equity.

In accordance with the other chapters in this volume, we encourage higher education leaders and policymakers to consider resilience throughout the pathway to STEM PhDs. For example, increasing resilience in the STEM PhD pathway would necessitate interventions to encourage more Latinas who earn undergraduate degrees in STEM to pursue PhDs in STEM as opposed to non-STEM fields. Latinas are part of one of the largest and fastest-growing minority populations in the United States, and if we do not graduate more US-born Latinas from STEM PhD programs, then the United States will fail to live up to being the land of opportunity and an economic leader throughout the twenty-first century.

# Note

1. We use the term "Latinas" even though the NSF data that were analyzed in prior research described PhD earners as "Hispanic" (e.g., Solórzano, 1994; Solórzano, 1995; Thurgood, Golladay, & Hill, 2006; Watford et al., 2006).

# References

Alvarez, R. R. Jr. (1986). The Lemon Grove incident. *The Journal of San Diego History, 32*(2), 116–135.

Brown v. Board of Education of Topeka (1954). 347 U.S. 483, 74 S.Ct. 686, 98 L.Ed.873.

Colby, S. L., & Ortman, J. M. (2015). *Projections of the size and composition of the U.S. population: 2014 to 2060 (Report # P25–1143)*. Washington, DC: U.S. Census Bureau. Retrieved from www.census.gov/content/dam/Census/library/publications/2015/demo/p25-1143.pdf

Congressional Research Service (2012). *Science, technology, engineering, and mathematics (STEM) education: A primer*. CRS Report for Congress. Retrieved from https://fas.org/sgp/crs/misc/R42642.pdf

Contreras, F. E., & Gándara, P. (2006). The Latina/o Ph.D. pipeline: A case of historical and contemporary under-representation. In J. Castellanos, A. M. Gloria, & M. Kamimura (Eds.), *The Latina/o pathway to the Ph.D.: Abriendo caminos* (pp. 91–111). Sterling, VA: Stylus.

Espinosa, L. (2011). Pipelines and pathways: Women of color in undergraduate STEM majors and the college experiences that contribute to persistence. *Harvard Educational Review, 81*(2), 209–241.

Estrada, M., Burnett, M., Campbell, A. G., Campbell, P. B., Denetclaw, W. F., Gutiérrez, C. G., . . . Zavala, M. (2016). Improving underrepresented minority student persistence in STEM. *CBE—Life Sciences Education, 15*(5), 1–10.

Fairlie, R. W., Hoffmann, F., & Oreopoulos, P. (2014). A community college instructor like me: Race and ethnicity interactions in the classroom. *American Economic Review, 104*(8), 2567–2591.

Fayer, S., Lacey, A., & Watson, A. (2017). *STEM occupations: Past, present, and future*. Washington, DC: U.S. Bureau of Labor Statistics. Retrieved from www.bls.gov/spotlight/2017/science-technology-engineering-and-mathematics-stem-occupaions-past-present-and-future/pdf/science-technology-engineering-and-mathematics-stem-occupations-past-present-and-future.pdf

Fernandez, F. (2018). Understanding the (sub)baccalaureate origins of Latina/o doctorates in education, humanities, and social science fields. *Hispanic Journal of Behavioral Sciences, 40*(2), 115–133.

Fernandez, F. (2019). What's money got to do with it? An examination of the relationships between sources of financial support and the post-graduation career plans of Latina and Latino doctoral students in the social sciences. *The Review of Higher Education, 43*(1), 143–168.

Fernandez, F. (2020). Where do Latinas and Latinos earn social science doctorates? *Education Policy Analysis Archives, 28*(97), 1–25.

Fernandez, F., & Baker, D. P. (2017). Science production in the United States: An unexpected synergy between mass higher education and the super research university. In J. J. W. Powell, D. P. Baker, & F. Fernandez (Eds.), *The century of science: The global triumph of the research university* (Volume 33: International Perspectives on Education and Society Series) (pp. 85–111). Bingley, United Kingdom: Emerald Publishing.

Fernandez, F., Baker, D. P., Fu, Y. C., Muñoz, I. G., & Ford, K. S. (April 2017). *The secret of American science capacity: Mass higher education and STEM Ph.D. production over the 20th century.* Research paper presented at the American Educational Research Association Annual Meeting, San Antonio, TX.

Garcia, M. (2006). The brown diamond: A Latina in the sciences. In J. Castellanos, A. M. Gloria, & M. Kamimura (Eds.), *The Latina/o pathway to the Ph.D.: Abriendo caminos* (pp. 243–253). Sterling, VA: Stylus.

Godfrey, P. C. (2008). The "other White": Mexican Americans and the impotency of whiteness in the segregation and desegregation of Texan public schools. *Equity & Excellence in Education, 41*(2), 247–261.

Gonzalez, K. (2015). *Graduate degree aspirations of Latino undergraduate engineering students at a research institution* (Unpublished master's thesis). California State University at Sacramento, Sacramento, CA.

Griffin, K. A., Muñiz, M. M., & Espinosa, L. (2012). The influence of campus racial climate on diversity in graduate education. *The Review of Higher Education, 35*(4), 535–566.

Grimes, M. D., & Morris, J. M. (1997). *Caught in the middle: Contradictions in the lives of sociologists from working-class backgrounds.* Westport, CT: Praeger Publishers.

Hernandez v. Driscoll Consolidated Independent School District (1957). Civ. Action. 1348, 2 Race Rel. L. Rptr. 329 (S.D. Texas).

Malcom, L. E., & Dowd, A. C. (2012). The impact of undergraduate debt on the graduate school enrollment of STEM baccalaureates. *The Review of Higher Education, 35*(2), 265–305.

Malcom, L. E., Dowd, A. C., & Yu, T. (2010). *Tapping HSI-STEM funds to improve Latina and Latino Access to the STEM professions.* Los Angeles, CA: University of Southern California.

Martínez, A., & Gayfield, A. (2019). *The intersectionality of sex, race, and hispanic origin in the STEM workforce (SEHSD working paper #2018–27).* Washington, DC: U.S. Census Bureau Social, Economic, and Housing Statistics Division. Retrieved from www.census.gov/content/dam/Census/library/working-papers/2019/demo/sehsd-wp2018-27.pdf

Millett, C. M., & Nettles, M. T. (2006). Expanding and cultivating the Hispanic STEM doctoral workforce: Research on doctoral student experiences. *Journal of Hispanic Higher Education, 5*(3), 258–287.

National Center for Science and Engineering Statistics (2011). *Data tables* (Table 4–3. S&E and S&E technologies associate's degrees awarded, by sex, citizenship, race/ethnicity, and field: 2010). Retrieved from https://ncses.nsf.gov/pubs/nsf19304/prior-releases

National Center for Science and Engineering Statistics (2019a). *Data tables* (Table 5–4: Bachelor's degrees awarded to women, by field, citizenship, ethnicity, and race: 2006–16). Retrieved from https://ncses.nsf.gov/pubs/nsf19304/data

National Center for Science and Engineering Statistics (2019b). *Data tables* (Table 6–4: Master's degrees awarded to women, by field, citizenship, ethnicity, and race: 2006–16). Retrieved from https://ncses.nsf.gov/pubs/nsf19304/data

National Opinion Research Center (NORC) at the University of Chicago (2012). *Methodology report for the Survey of Earned Doctorates academic year 2011.* Chicago, IL: NORC.

National Science Board (2018). *2018 science & engineering indicators*. Retrieved from www.nsf.gov/statistics/2018/nsb20181/assets/nsb20181.pdf

National Science Foundation (2017). *Data tables* (Table 19: Doctorate recipients, by ethnicity, race, and citizenship status: 2006–16). Retrieved from www.nsf.gov/statis tics/2018/nsf18304/data.cfm

Parsons, T. (1946). The science legislation and the role of the social sciences. *American Sociological Review, 11,* 653–666.

Parsons, T., & Platt, G. M. (1968). Considerations on the American academic system. *Minerva, 6,* 497–523.

Pérez Huber, L., Huidor, O., Malagón, M. C., Sánchez, G., & Solórzano, D. G. (2006). *Falling through the cracks: Critical transitions in the Latina/o educational pipeline* (Chicano Studies Research Center Report No. 7). Retrieved from http://files.eric.ed.gov/full text/ED493397.pdf

Posselt, J. R. (2016). *Inside graduate admissions: Merit, diversity, and faculty gatekeeping.* Cambridge, MA: Harvard University Press.

Posselt, J. R. (2018). Normalizing struggle: Dimensions of faculty support for doctoral students and implications for persistence and well-being. *The Journal of Higher Education, 89*(6), 988–1013.

Powell, J. J. W., Baker, D. P., & Fernandez, F. (Eds.). (2017). *The century of science: The global triumph of the research university* (Volume 33: International Perspectives on Education and Society Series). Bingley, United Kingdom: Emerald Publishing.

Rendón, L. I., Nora, A., Bledsoe, R., & Kanagala, V. (2019). *Científicos Latinxs: The untold story of underserved student success in STEM fields of study.* San Antonio, TX: Center for Research and Policy in Education, The University of Texas at San Antonio.

Ro, H. K., & Knight, D. B. (2016). Gender differences in learning outcomes from the college experiences of engineering students. *Journal of Engineering Education, 105*(3), 478–507.

Ro, H. K., Knight, D. B., & Loya, K. I. (2016). Exploring the moderating effects of race and ethnicity on the relationship between curricular and classroom experiences and learning outcomes in engineering. *Journal of Women and Minorities in Science and Engineering, 22*(2).

Ro, H. K., Lattuca, L. R., & Alcott, B. (2017). Who goes to graduate school? Engineers' math proficiency, college experience, and self-assessment of skills. *Journal of Engineering Education, 106*(1), 98–122.

Ro, H. K., & Loya, K. I. (2015). The effect of gender and race intersectionality on student learning outcomes in engineering. *The Review of Higher Education, 38*(3), 359–396.

Ruiz, E. C. (2013). Motivating Latina doctoral students in STEM disciplines. *New Directions for Higher Education, 2013*(163), 35–42.

Slay, K. E., Reyes, K. A., & Posselt, J. R. (2019). Bait and switch: Representation, climate, and tensions of diversity work in graduate education. *The Review of Higher Education, 42*(5), 255–286.

Solórzano, D. G. (1994). The baccalaureate origins of Chicana and Chicano doctorates in the physical, life, and engineering sciences: 1980–1990. *Journal of Women and Minorities in Science and Engineering, 1*(4), 253–272.

Solórzano, D. G. (1995). The baccalaureate origins of Chicana and Chicano doctorates in the social sciences. *Hispanic Journal of Behavioral Sciences, 17*(1), 3–32.

Thurgood, L., Golladay, M. J., & Hill, S. T. (2006). *U.S. doctorates in the 20th century*. Arlington, VA: National Science Foundation. Retrieved from https://immagic.com/eLibrary/ARCHIVES/GENERAL/US_NSF/N060630T.pdf

Torres, V., & Hernandez, E. (2009). Influence of an identified advisor/mentor on urban Latino students' college experience. *Journal of College Student Retention: Research, Theory & Practice, 11*(1), 141–160.

Turner, C. S. V., González, J. C., & Wood, J. L. (2008). Faculty of color in academe: What 20 years of literature tells us. *Journal of Diversity in Higher Education*, 1, 139–168.

U.S. Bureau of Labor Statistics (2017). *Spotlight on statistics*. Retrieved from www.bls.gov/spotlight/2017/science-technology-engineering-and-mathematics-stem-occupations-past-present-and-future/home.htm

U.S. Census Bureau (2019). *About the American community survey*. Retrieved from www.census.gov/programs-surveys/acs/about.html

Warnock, D. M., & Appel, S. (2012). Learning the unwritten rules: Working class students in graduate school. *Innovative Higher Education, 37*(4), 307–321.

Watford, T., Rivas, M., Burciaga, R., & Solórzano, D. G. (2006). Latinas and the doctorate: The "status" of attainment and experiences from the margin. In J. Castellanos, A. M. Gloria, & M. Kamimura (Eds.), *The Latina/o pathway to the Ph.D.: Abriendo caminos* (pp. 112–133). Sterling, VA: Stylus.

Woodrow Wilson National Fellowship Foundation (2005). *Diversity and the Ph.D.: A review of efforts to broaden race & ethnicity in U.S. doctoral education*. Princeton, NJ: Author.

# 4 "Cuida Tu Casa Y Deja La Ajena"

## Focusing on Retention as a Self-Perpetuating Engine for Recruiting Latina Faculty in STEM

*Aurora Kamimura*

Much like the *consejos* (advice) that I would often receive from my *abuelitas* (grandmothers) and *tías* (aunts) when I would seek their advice, the *consejo* in this chapter centers on a *dicho* (saying). Often when we are struggling, we look for external answers and solutions. However, as my *familia* (family) always reminds me, our greatest strength comes from within, and so do our best solutions. Therefore, while we can learn from the achievement of others, one of the most pivotal strategies for embracing the challenge of recruiting, hiring, and retaining Latina faculty, especially within STEM, requires that we "*cuida tu casa y deja la ajena*" (take care of your own "home" and let others be). In similar ways to the internal strength that dwells within my *familia*, STEM departments focusing internally will enhance the organizational resilience upon which their *casa* will thrive.

## National Landscape: The State of Latina STEM Faculty in the United States

As indicated in previous chapters, the demographics of our nation have been rapidly shifting. One of the largest noted shifts has been within the Latinx community, with remarkable growth from 4.5% of the national population in 1970 to 16.3% in 2010 (Núñez, Hurtado, & Calderón Galdeano, 2015), and now with a projected leap to 30% of US residents by 2050 (Saenz, 2010). While this estimation has been anticipated by the educational community and the nation at large, the proportions of Latinx students pursuing and achieving bachelor's degrees after six years of first-time college enrollment is still only at approximately 7% (Gándara & Contreras, 2009). As has been noted in previous chapters, more Latinx students are entering college, but many are not making it out successfully.

While the funnel narrows significantly with such small proportions of Latinx students attaining a bachelor's degree, if we truly aim to understand how to infuse the professoriate, we must consider the critical junction between the doctorate and academic positions (i.e., assistant professor, associate professor, professor, instructor). In 2013, according to the National Science Foundation, of all US doctoral graduates (N=52,760) only 4% identified as Hispanic/Latinx. When considering gender, the data indicate that 37.8% of all science, technology, engineering, and math (STEM) doctorates were attained by females,[1] but only 5% of all female doctoral recipients (n=1,225) were Latina (NSF, 2013). Therefore, we begin to see how the intersectionality of race/ethnicity and gender truly narrows the pool of those Latinas eligible for faculty positions upon graduation.

Nationally, the 2013 landscape on opportunities for full-time faculty positions (tenure and tenure track) shows that females hold 26.9% of these positions, while faculty of color only hold 9.2% of equivalent positions (NSF, 2014). Subsequently, Latinx faculty are extremely underrepresented in academia by occupying merely 2.3% of all full-time faculty positions in all disciplines (NSF, 2014). The inequities in the professoriate range by discipline; however, they are collectively most pronounced in STEM. According to the National Science Foundation, in 2013 universities and 4-year colleges hired 354,000 faculty (i.e., teaching, research, and adjunct faculty) in STEM programs. Of these faculty, 34% were women, and only 9.6% of these positions were occupied by faculty of color. According to the National Center for Education Statistics, Latinx faculty comprised a mere 1.3% of all STEM faculty in 2003 (NCES, 2008).

Furthermore, when these data are disaggregated by position, the inequities increase for part-time faculty. For instance, 42% of women in academia are part-time faculty, which represent the least stable positions in the professoriate (NSF, 2013). Moreover, in STEM, women are about 6% less likely to be hired into a faculty position than in other academic fields, while STEM faculty of color are hired at a nearly 10% lower rate than in other disciplinary fields (NSF, 2013). Therefore, to say that we have work to do is an understatement. Although the current national landscape looks bleak for Latina STEM faculty, there are glimmers of hope in various departments and programs across the United States that are working intentionally to open the doors to their *casas*. The thematic findings presented in the latter half of this chapter shed light on promising strategies for recruiting and hiring women faculty and faculty of color in STEM, with deep implications for Latina STEM faculty.

### Understanding the Gatekeeper: The Process Between the Doctorate and the Professoriate

In an attempt to better understand the persisting inequities in the professoriate, several rationales have been considered in the extant literature on minoritized faculty, including the shallow pool of diverse candidates for faculty positions (Kayes, 2006; Turner, 2002), unconscious bias impacting departmental processes including hiring and promotion and tenure (Smith,

2000; Stewart, Malley, & LaVaque-Manty, 2010; Turner, 2002; Turner, González, & Wood, 2008), and attrition of minoritized faculty (Moreno, Smith, Clayton-Pedersen, Parker, & Teraguchi, 2006). However, various studies have proven that the lack of qualified diversified applicants is indeed a myth, and rather an excuse for biases at play (Olivas, 1988; Smith, 2000; Turner, Myers, & Creswell, 1999). Therefore, focusing on the experiences of minoritized faculty (i.e., women and faculty of color) will provide the strongest path toward a more equitable future.

### Structural Gatekeepers: The Processes and Practices

As identified previously, unconscious bias and lack of transparency in structural processes and practices—such as recruitment and hiring of faculty and promotion and tenure of faculty—can often lead to the greatest barriers that faculty have to overcome, especially minoritized faculty (Smith, 2000; Stewart et al., 2010; Turner, 2002; Turner et al., 2008).

Stanley (2006) argued that challenges stemming from negative student evaluations from teaching on contentious issues relating to diversity often negatively impact a faculty member of color's promotion and tenure process. Furthermore, Stanley (2006) stipulates that faculty of color lack mentoring that is necessary in learning the unspoken *rules of the game*. Issues at the intersection of collegiality and identity are often misinterpreted, and faculty of color are typically not protected from overloaded service requests (formal and informal). Finally, faculty of color often have to manage racism that persists as a result of working in a primarily White environment. Taken together, these challenges deeply impact the performance of faculty of color during the promotion and tenure process. Without structural components in place to equalize the playing field, most faculty of color will continue to struggle to succeed through the tenure track and promotion process.

However, while retention and promotion of women faculty and faculty of color are a critical foundation to sustaining a diverse faculty, studies have identified recruitment and hiring processes as the primary gatekeeper en route from the doctorate to the professoriate (Bilimoria & Buch, 2010; Olivas, 1988). Insights from research and practice have indicated that various stages in the recruitment and hiring process require increased training to minimize bias. Several stages that have been the focal points of numerous articles and handbooks include intentionality of leadership prior to the start of the formal search process; the composition of the search committee; job postings (as a leveraging point to create a more diverse applicant pool); broadening the scope of merit in the application review process; and training to minimize unconscious bias throughout the search process (Alger, 2000; Turner, 2002). Therefore, if the search process is the main point of entry into academia, the literature clearly indicates that a revamping of the multi-layered stages throughout the process is foundational for stronger success in recruiting and hiring more minoritized faculty.

## Environmental Gatekeepers: "Chilly Climates" and "Revolving Doors"

While significant portions of the higher-education literature on STEM faculty highlight persisting compositional inequalities (Burelli, 2008; Turner et al., 2008), the marginalizing climate and experiences that women faculty and faculty of color manage also have been studied as influencing high attrition. In particular, Turner and colleagues (1999) pointed to the "chilly climate" that many faculty of color experience while aiming to be successful on the tenure clock, overcoming additional barriers including "being told they did not fit the 'profile' of someone being promoted," and performing while "being in the spotlight" (pp. 41–42). Similarly, other scholars have argued that the microaggressive behavior that faculty of color experience on a daily basis (Griffin, Pifer, Humphrey, & Hazelwood, 2011; Reyes & Halcon, 1988), on top of the racism and cultural taxation experienced primarily by faculty of color through heavier service responsibilities and informal mentoring roles to Students of Color (Griffin, Bennett, & Harris, 2011), all contribute to the chilly climate that eventually leads to faculty of color leaving a given department. Greene and Stockard (2010) found parallel climate issues, specifically in chemistry, for women faculty concerned with the climate being "problematic and less than welcoming" (p. 381) and receiving inequitable resources to their male counterparts such as salaries, workload, workspace, and research recognition.

Even departments that are intentionally working toward recruiting and hiring more faculty of color may suffer from an additional issue that perpetuates the inequity in the professoriate—the *revolving door syndrome* (Moreno et al., 2006). With the revolving door syndrome, the attrition of faculty of color is so significant that in spite of focused efforts to hire more faculty of color, the numbers remain the same. Basically, as the department works to bring in a new faculty of color, they also lose one through the revolving door. While the "chilly climate" contributes significantly to the attrition of faculty of color (Turner et al., 1999), the revolving door syndrome deeply compounds the issue and deepens inequity, especially for faculty of color.

### Passing by the Gatekeeper: Entering the Casa

As the higher-education literature on minoritized faculty highlights in the experiences in the search process, as isolated faculty of color in their departments, and though the promotion and tenure process are not overwhelmingly positive, there are glimmers of hope indicating where change might begin. Practitioners and campus administrators have been advised to consider faculty recruitment as a daily routine (Olivas, 1988) and have been urged to consider focusing on their departmental climate as a self-perpetuating engine for recruitment of minoritized faculty (Light, 1994). Light's (1994) work recognized the importance of developing a healthy environment where faculty of color could succeed and thrive, and he argued that prospective faculty of color would look for these signals of success as indicators of the support they would receive once hired. In many ways, focusing internally would slow the revolving door syndrome (Moreno et al., 2006) significantly, and would begin to reverse the impact of the "chilly climate" (Turner et al., 1999) that drives many faculty of color away from departments or academia altogether.

Although much of the literature on minoritized faculty has not considered Latinx faculty or Latinas specifically, the concept of *familia* and familial support has been studied to be at the core of *Latinidad* in education. More specifically, Gloria and colleagues (2005) underscore the power of familial support as an indicator of success for Latinx students; however, it is clear that familial support continues to be a core cultural value for Latinx faculty as well (Yosso, 2005). Understanding the ways in which *familia* and familial support could look for Latina faculty (especially in STEM, where they are so deeply underrepresented) could lead to the discovery of new recruitment and retention strategies.

Moreover, for many faculty of color, having a connectedness to "home" and to community strengthens their sense of belonging and investment in their careers (Reyes, Carales, & Sansone, 2020). In their autoethnography of their faculty search and recruitment process, Reyes and colleagues (2020) discovered that their desire to be "homebound" and to have the opportunity to serve their community became a pivotal decision-maker in the job search process. As faculty of color, they discussed how important being close to "home" and to their community was to their perceived success as tenure-track faculty. Similar to the concept of *familia*, "home" can vary in physical characteristics, and one could beg to say even in location; however, for Latina faculty who may thrive from the connectedness of *familia*, feeling a sense of "home" in their department and a connectedness to the community could become a game-changer in their success as well. Therefore, as campus leaders are urged to focus on their organizational culture (Light, 1994), considering strategies for developing a sense of *familia* and "home" could instill a sense of belonging that would help Latina faculty in STEM be more resilient and included.

## A Campus Strategy for Success: Developing the Self-Perpetuating Engine

As the literature has noted, there is so much work to be done before reaching a place of equity within the STEM professoriate, especially for Latinas. However, there are glimmers of hope that can be gleaned from STEM departments that are intentionally working toward recruiting more women faculty and faculty of color. Presented here are analyses from a five-unit comparative case study analysis at Midwestern University (MU)[2], which indicate that recruiting and hiring more women faculty and faculty of color in disciplines where they are severely minoritized requires intentionality and multifaceted approaches (Kamimura, 2019[3]). Three of the departments—astronomical sciences (AS), chemistry, and physics—argue that focusing on the internal organizational climate can be the best recruitment strategy. In doing so, three specific recommendations arose: intentionally promoting inclusion, empowering the minoritized, and signaling support structures.

### Midwestern University Context

MU shares many institutional characteristics with other large, public, very high research-intensive campuses in the United States. However, its uniqueness stems from a long history and relentless commitment to social justice. MU currently matriculates more than 40,000 students (combination of undergraduate and graduate) in close to 20 schools and colleges, for one or more of approximately 300 degree programs. Of the undergraduates enrolled at MU, more than 50% are female, and less than 15% are Students of Color. Conversely, only about 10% of all faculty employed on campus are people of color. Furthermore, MU is located in a state that is legally restricted from using social identities in the hiring process for university employees. Thus, while MU experiences a gap in representation between Students of Color and faculty of color, and they are in a restrictive state, they remain loyal to their social justice history in aiming toward equity in the professoriate.

Three of MU's departments in particular—astronomical sciences (AS), chemistry, and physics—were selected as exemplar departments to analyze based on their investment in diversity, equity, and inclusion efforts on campus (Merriam, 2009). However, each of these departments has a unique history and rationale for investing in these efforts. Most notably, AS is the department with the youngest organizational history, yet it was experiencing some turmoil about ten years ago with a significant number of departures, most notably by women faculty. This attrition drew national attention and ultimately negatively impacted their reputation. Chemistry was contending with a "head-hanging" history, according to one of their long-standing faculty, after struggling through a gender equity lawsuit and recognition that the climate for female faculty was feeding a revolving door. Physics, on the other hand, was not contending with a tumultuous history; rather, it was struggling with high levels of internal resistance to change. Therefore, the physics department had to lean heavily on the college's leadership to pave the way, thus creating a culture of compliance for the early years when faculty were being pressed by the college and MU to increase diversity in their faculty. Regardless of the differing reasons for implementing strategies to recruit and hire more women faculty and faculty of color, all three departments have been intentionally invested in efforts and realized the power of creating and sustaining a positive departmental climate as the best recruitment (and retention) strategy. Insights from key faculty and leaders in each of these departments point to the following three recommendations to actively create and sustain a positive environment.

### Intentionally Promoting Inclusion: Structural Environments

Through the study, intentionality was central to the findings. In this case, intentionality was foundational in promoting inclusion within departmental cultures. As such, chemistry and physics infused more transparency into their policies and practices, while AS focused intently on many structural markers. It was evident that changes to the organizational structures were needed to allow for psychosocial factors in the organizational climate to also change.

#### Transparency and Review of Policies

The history of the departments discussed in this chapter—AS, chemistry, and physics—each pointed to a deeply needed review of current policies and practices and an infusion of more transparency. Chemistry and physics showed how transparency and revision to their policies impacted their success in recruiting more women faculty and faculty of color. In physics, specifically, James (a previous chair of the department) shared how opening the scope of research area in their recruitment process lead to a more diverse pool. He commented:

> One of the things we've done here . . . was to do broad-based searches open to various fields. If, for example, you said, I want someone who is going to look at capillary data—a particular satellite telescope—then now, you've limited it to five people in the world. Whereas if you say, I'm going to look broadly across all fields of physics, now you've opened it up to hundreds of potential applicants. . . . What that does is you've greatly expanded your potential pool of candidates. The disadvantage is the faculty who are advocating for a hire in their particular field. . . . [As chair,] you say, well, maybe the best person isn't in that field this year.

Repeatedly, each of the four faculty and leaders interviewed in physics commented on the power of having revamped their application process to a broadened search as the solution to creating a more diverse pool of applicants and thus higher probability of a woman and/or faculty of color hire. As frequently noted, this change in policy and practice came with high levels of resistance early on from many faculty members. However, as success was garnered one search at a time, physics faculty eventually resisted less, and now broad-based searches are expected for any hiring process that takes place in the department, inclusive of postdoctoral researchers and research assistants.

In a department like chemistry that was rebuilding from a damaged history reflective of a chilly climate (Turner et al., 1999) for women faculty, the road toward recovery was much steeper. Therefore, they engaged in significant reflective work to understand that transparency and clarity in the tenure and promotion process were lacking, thus not allowing for many (especially women faculty) to feel successful. As Diana, the current chair, entered the department, she quickly learned about the current practices at the time. She recalled:

> The department . . . had secret mentoring committees—secret evaluation committees. . . . They wrote these reports every year about how they were doing, but you didn't know who was on the committee. So, I remember I was here and Luke . . . came into my office and said, "Would you support our proposal to get rid of the secret committees." I said, "What do you mean secret committees? What secret committees?" So I said yes.

The secretive process was reviewed and revamped. As Luke, a long-standing faculty in chemistry, shared:

> [T]he department completely changed its tenure and promotion policies on the order of 15 years ago to a much more transparent, much more open way that I think is better for anybody. It wasn't designed in a gender-specific way, but I think a lot of the changes were the ones that helped the women coming through quite a bit.

After speaking with several faculty and leaders in the department, it became evident that the review of these policies was just the beginning of a longer review process that ensued over the following years in an effort to infuse more transparency and clarity. As the policy review process continued, the department's reputation and organizational climate began to drastically change for the better. Judy, a faculty diversity resource at MU, commented:

> I think that it became a place that was really well known to be good. It was the opposite of their previous reputation. They literally turned around a negative reputation as a place to be as a woman to a positive one. It became a zero tolerance for sexual harassment place, which had been the opposite.

Each of the faculty and leaders in chemistry spoke to the positive results in recruitment, hiring, and retention to which these policy changes led. Ensuring that they would never return to the "head-hanging" place where they once were, each faculty committed informally to ensuring that they contributed to creating an inclusive and welcoming environment for all, as well as continuing to be committed to the revision of policies to ensure a positive organizational climate.

*Gender-Neutral Restrooms*

Compared to the structural changes that physics and chemistry implemented in their recruitment and promotion processes, AS focused more on physical areas that would signal inclusion. Marc, the chair of AS, proudly pointed out several of the physical signals that have marked their progress toward inclusiveness as a department, including departmental stickers and T-shirts with diversity-related slogans; newer posters and wall décor in the hallways reflecting the diversity of the field; and most markedly, the gender-neutral restrooms that they fought to have on their floor. It is worth noting that while all of the departments were invested in creating change within their departments, AS is the only one of the five departments interviewed that fought for gender equity in their restrooms. Inclusive, in several of the departments included in this study, finding a restroom was often a challenge; however, AS ensured that the restroom designated as a gender-neutral restroom was central to the floor and accessible from both sides of the hallways. While AS did not have any faculty or staff that identified as gender non-conforming, they wanted to ensure that this option would be available to anyone who would prefer to use a gender-neutral restroom without asking. This was a significant structural marker of inclusion.

**Empowering the Minoritized: Enhancing Resilience**

While women and faculty of color are minoritized within STEM disciplines, chemistry and AS show us strategies to empower these faculty in explicit and implicit ways. In efforts to develop warmer departmental cultures, chemistry and AS concurrently developed the foundation for their organizational resilience at the group level. As scholars (Sutcliffe & Vogus, 2003) have argued, organizational resilience at the group level requires three foundational mechanisms: (1) accumulation of knowledge; (2) diversity of group's membership; and (3) experiential diversity within the group's membership. It is through the enhancement of these mechanisms that a group begins to recognize their own strength and resilience that is necessary to weather challenges when they arise. Therefore, the chairs of the chemistry and AS both focused on ensuring that diversity within each of their units was integrated, heard, and visible, as much as possible.

*Building Critical Mass*

Many of us may already know that developing a critical mass of minoritized individuals within an organization relieves the feeling of isolation that frequently leads to attrition. However, chemistry reinforces how the concept of critical mass remains relevant, and even more so critical, for the success of minoritized faculty. Steve's (a long-standing chemistry faculty member) recollection underscored this value:

> You've got to have more than one. It is very clear. The literature is right. That first person becomes the exemplar for everything and the person who carries all the burden and weight of being the representative. And it isn't until you build the variation, the diversity, within that identifiable subpopulation, so that everybody else can learn, that there isn't just a monotheistic view that comes with that. And it is remarkable how it works. I've seen it work. Then you get to something called a critical mass.

Again, while this may seem so basic to some, for other departments, creating this critical mass is much more challenging. Thus, this work requires intentional effort and focus to reach a critical mass. And although "critical mass" is an elusive term that is often challenging to quantify, for a faculty member feeling the isolation of being "the only one," concerted efforts to recruit a critical mass may mean the difference between success and departure.

We may ask, "Well, how does a department specifically go about developing a critical mass?" Chemistry faculty member Luke shared a prime example. He frequently referred to Nicki's hire as one of the department's greatest successes. As such, he spoke of how hiring Nicki, whom he characterized as "arguably the top chemist in her field in her generation," was a windfall for the department. Luke pointed to Nicki as the first of many women chemists who later joined, and pointed to this memory as to why she chose to join the MU chemistry department:

> [S]he came because she perceived . . . this changing culture and a culture that was going to allow her to do her science and live her life and be happiest in her career. So, she came. So, Diana coming was a huge watershed moment. Nicki coming was a huge watershed moment for the department.

Luke and several other chemistry faculty members distinctly pointed to Nicki's arrival as the first of many women to join the department and successfully attain tenure. They recognized how with each woman faculty member who was recruited and chose to join them, they grew stronger, and more women wanted to join because they perceived a sense of support and attainable success. Therefore, not only did this critical mass relieve the sense of isolation, it also signaled a sense of possible success, much like their women colleagues who were already there.

*Voice Is Power: Supreme Court Rules*

AS was quite proud of the strategy that they instilled to empower the assistant professors in the department, which they quickly recognized was where the "diversity" resided. Therefore, Marc (the chair of AS) excitedly piloted a strategy he heard about:

> With contentious topics, I want to minimize the strong voices that are always trying to dominate the conversation. So, I structure the conversation using what I call "Supreme Court rules." And I've been telling everybody about this because I think it is fantastic, where the Supreme Court has a rule that when they first discuss a case that they go around the table. Each person gets to state their opinions—before Scalia at the time would interrupt them. What I do is, okay, we're going to go around the table. And everybody gets their say. The very useful effect is that the assistant professors would contribute to the conversation because they wouldn't get talked down. And so I empowered the diversity in the department because diversity represents the bottom end and not the top end of the age distribution.

Marc eagerly shared this strategy because he has had great success with it. He realized that implementing the Supreme Court rules allowed for all voices to be heard, equitably, and allowed for all faculty to feel empowered in all conversations—especially the most challenging ones. Therefore, not only did this create a stronger sense of inclusivity; Marc also noted that it changed the nature of the content discussed, as diverse perspectives were not equitably shared and considered in decision-making processes.

*"Children in the Hallways."*

As a final strategy learned from chemistry on how to empower minoritized faculty, they focused intently on changing the climate and culture of the department. As mentioned previously, this began with changes to structures and policies, but it also evolved into some implicit spaces that also signaled inclusivity of varied social identities. Diana, the department chair, reflected on what she believed signals inclusivity:

> I think almost all of our faculty believe that we should have pictures of our family out and so little cues. And at our graduate recruiting weekends frequently people bring their kids, so [students] see kids running around. The male faculty can bring their kids too, but they definitely see that this isn't a place where, you know, people believe you should never have children. And so, I think it is not explicit clues but some of these implicit clues.

For a department that struggled with issues of sexual harassment and chilly climate (Turner et al., 1999) in prior years, the thought of being a "whole" intersectional self was never a consideration. The faculty and leadership in chemistry recognized that while some of the women faculty may have children, that this shift in family-friendly climate positively impacted men faculty with families as well. As they repeated often throughout the interviews, these practices and policies were not developed for one population in mind, but it definitely signaled a more inclusive environment.

Moreover, chemistry faculty explicitly signaled that they were a family-friendly department, and a department where faculty members enjoyed spending time with each other. One of the strategies that they explicitly planned for was during faculty recruitment visits. Steve, a long-standing faculty member, shared:

> So, when we are going to hold an event . . . we would like to be a family-friendly event for recruiting [therefore] I want to bring in 40 faculty members, their partners and their children.

He went on to share how they rent out local eateries or dessert shops in town for events and create fun activities for the faculty and their families to engage in together. Steve's face lit up as he shared the fun they have as a faculty, hosting events and welcoming potential faculty. As he shared, it is a moment for them as a faculty to gather and enjoy each other and their families, and it is also a nice moment for future colleagues to see how well they all get along. Steve spoke for a while about this strategy, as he has learned over the years the value of inviting future colleagues into a community and helping them see that they will be joining a collaborative department that enjoys each other as well. He ended by sharing that this is one of their most successful strategies, and also most enjoyable.

### Signaling Support Structures: Creating Familia

This final strategy is about the implicit mechanisms that the departments have in place to signal that support will be available once the future faculty arrive. All three of the departments recognized that embedding these signals not only creates its own recruiting engine for future faculty but also, importantly, are retention strategies for their current faculty. These are hugely impactful in stopping the "revolving door syndrome" (Moreno et al., 2006), and in enhancing the resilience (Sutcliffe & Vogus, 2003) necessary in creating a strong *familia*.

#### Leadership Mentoring of Assistant Professors

Physics honed in particularly on the valuable mentoring that the chair distinctly provided the tenure-track faculty. During the interviews, it was evident that mentoring was a foundational expectation for their culture. Marie, an associate professor in physics, described her department climate as "healthy," reflecting the challenge and support that exists, as well as balancing of expectations. Marie summed up her impressions of the climate:

> I feel fortunate that we have a really collegial department overall. . . . It is like a healthy department, which is easier said than done, if you don't have a healthy department. . . . When somebody comes for an interview . . . [during] most of their interactions they get a vibe that it's a sort of healthy department. And people aren't all being snide against each other and so all of that feeds into a friendlier sort of place. And I think that comes through, again, in the selection process on the search committee—helps a lot, and when people are interviewing.

She further elaborated regarding the expectations for mentoring and the ways in which the healthy climate positively impacts the faculty recruitment process:

> I think . . . that our chair values mentorship—mentorship by more senior faculty of junior faculty, mentorship of graduate students. And he values teaching a lot too, which I appreciated when I came here because I was coming from a national lab staff scientist position where I had only research responsibilities. And I consciously wanted to transition to a university position where I would have [both]. . . . Having a [healthy] culture and, in particular, probably I would imagine it comes up [with] the chair—during the candidate's interview when they meet with the chair. I would imagine between teaching and mentoring, the sense that our current chair values investing in young people's futures as a real investment. . . . Yeah, I think that helps [with recruitment].

It is clear from Marie's insights that mentoring plays a significant role in signaling to potential faculty colleagues that they will receive the support they need to be successful, and that they are supported to meet all of their expectations equitably: research, teaching, and service/mentoring. According to Marie, there was no denying that interviewees would not gather a sense of a healthy and collaborative environment during their campus visits based on their overall departmental climate.

*Informal Role Modeling: "If She Can Be Successful, So Can I."*

Chemistry, in particular, pointed to the success of their fellow women faculty as positive draws for other potential women faculty they were trying to recruit. There was a sense that the prospective faculty could see how successful (i.e., thus supported) the women faculty were and felt a sense of relief that they, too, could be successful if offered the position. Luke, a long-standing faculty member, asserted:

> I don't think you have to look too much further on a map. Nicki, Diana, you know, and others so we have this cohort. There are a couple things about it. They seem to be happy here. They're succeeding here. They're winning the biggest awards in the country here. And I think that's why we're doing so well with recruiting, frankly.

As evidenced, he recognizes the value of seeing happy women with balanced lives being stars in the department, while not driving themselves into the ground. It was evident that the resilience present in the department, as evidenced by these successful female scholars, could transcend to their experiences as future scholars in that unit. When asked if he thought that having women faculty influenced their ability to recruit more women faculty, here is what Luke shared:

> You know [one of our star female faculty] and I compete for students all the time. . . . The good news is that they [students] have actually said to me, "Well, I didn't choose to work with one of them, but I'm sure much happier to be at Midwestern with them here." . . . Again there is the role model idea or "Is this department going to be friendly to me as I work in it?" When you have strong people like that in the department you're not so worried about being ostracized for your gender because, how would it happen?

Faculty members such as Luke are pointing to the necessity of creating *familia* in the departments. Basically, by creating *familia*, they are ensuring that minoritized faculty do not feel isolated and can indeed be successful.

*Seamless Transitions: Postdoc to Tenure Track*

The departments implemented a practice to seamlessly transition postdoctoral scholars into tenure-track faculty positions in an intentional effort to signal departmental support. AS has most proudly and successfully implemented this strategy. While many of the other STEM departments at MU have recruited and hired postdoctoral fellows from the same programs, none except AS have made a tenure-track offer following the term of the fellowship to ensure continued success. However, to the AS chair, Marc, this is a critical signal in ensuring the future success of these fellows. Therefore, he works diligently to ensure that all faculty are included in the selection process and that there is a sense of collectivism regarding any offer that is made. Tracy, an assistant professor in AS, shared a brief insight into the new direction for post-doctoral hires: "[T]hey are actually treating the next round of [postdocs] as a faculty hire." Stella, an associate professor in AS, confirmed:

"[A]stronomical sciences is the only department where it's been a post-doc position definitely followed by faculty." Each of these faculty confirmed that the purpose behind this seamless process was to signal to the postdoctoral fellow that AS was deeply interested in them and, most importantly, invested in their long-term success. This is one of the strongest possible signals of support—written confirmation and mentorship in place, from beginning to end.

## *"Cuidando Nuestra Casa"*: Implications for Practice and Leadership

While none of these three departments studied specifically spoke to strategies for recruiting Latinas into STEM departments, the three strategies that arose are applicable for recruiting (and retaining) Latina faculty, based on the literature previously discussed. With a high desire for a sense of "home," and a deep value for *familia* for Latinxs, it is fair to believe that strategies for creating a positive departmental climate that can be perceived would be a definite draw for Latina faculty to STEM. Furthermore, these strategies are pivotal for enhancing the organizational resilience necessary for innovation and thriving to transpire. Many of us have *chosen familia* as pillars of our success in our educational pipeline, and it is reasonable to assume that having the opportunity to sense a climate where *chosen familia* can be identified would be critical for successful recruitment, hiring, and retention of Latina faculty in STEM, especially in departments where they may be the only one.

Therefore, as learned from the three departments studied (AS, chemistry, and physics), departments should work intentionally on fostering a *warm climate* as opposed to a chilly climate (Turner et al., 1999) for their faculty. This warm climate would stop the revolving door syndrome (Moreno et al., 2006) and signal to prospective faculty that this could be a place where they may also be successful. Thus, the following three recommendations are highly encouraged for STEM department chairs working to recruit and hire more Latina faculty:

1. Intentionally promote inclusion on a daily basis. This requires a thorough review of current policies and practices in place, most prominently for hiring and tenure and promotion. It also requires the critical examination of structural signals that can be altered or added, to signal that *all* are welcome.
2. Empower the minoritized by enhancing resilience. This requires critical reflection regarding power structures within the departments in order to understand where power lies, and who holds the most powerful voices. Find strategies to bring more equity and integration of all voices. Also, increase implicit and explicit signals throughout locations and events that signal work–life balance and the ability to be an

intersectional scholar—with multiple identities, such as being a Latina woman in STEM.
3. Clearly signal support structures that are evident during the recruitment and hiring process. This requires an examination of processes and individuals involved in the recruitment and hiring process and finding strategies to signal that *chosen familia* can be found in the department, collaborative departments, or the university community. With *familia* as a central value to *Latinidad*, ensuring that community and support are authentically embedded throughout the structures of the department is critical.

Thus, success in hiring Latina faculty into STEM departments is achievable. It is a matter of focusing internally on the organizational climate of the department and allowing that *warm climate* to serve as the self-perpetuating engine for recruitment, and as a foundation for the organizational resilience of their *casa*.

## Notes

1. Although gender is fluid and non-binary (Butler, 1990; Linstead & Pullen, 2006), unfortunately national data are gathered binary measures. Thus, the data reported in this chapter align with these national reporting practices.
2. Henceforth, all names (inclusive of campus, departments/units, participants, programs, etc.) are reported via pseudonyms in an effort to maintain confidentiality.
3. For further details on the full study's research design and methodology, please see *"Untying Our Hands": A Mixed Methods Study of Strengths-Based Approaches to Faculty Recruitment and Hiring Practices in STEM for Equity* (Kamimura, 2019). The full study provides details regarding participant selection, coding processes, etc.

## References

Alger, J. R. (2000). How to recruit and promote minority faculty: Start by playing fair. *Black Issues in Higher Education, 17*(20), 160–161.
Bilimoria, D., & Buch, K. K. (2010). The search is on: Engendering faculty diversity through more effective search and recruitment. *Change: The Magazine of Higher Learning, 42*(4), 27–32.
Burelli, J. (2008). Thirty-three years of women in S&E faculty positions. *Infobrief, Science Resources Statistics, National Science Foundation,* 08–308.
Butler, J. (1990). *Gender trouble: Feminism and the subversion of identity.* New York, NY: Routledge.
Gándara, P. C., & Contreras, F. (2009). *The Latino education crisis: The consequences of failed social policies.* Cambridge, MA: Harvard University Press.
Gloria, A. M., Castellanos, J., Lopez, A. G., & Rosales, R. (2005). An examination of academic nonpersistence decisions of Latino undergraduates. *Hispanic Journal of Behavioral Sciences, 27*(2), 202–223.
Greene, J., & Stockard, J. (2010). Is the academic climate chilly? The views of women academic chemists. *Journal of Chemical Education, 84*(4), 381–385.

Griffin, K. A., Bennett, J. C., & Harris, J. (2011). Analyzing gender differences in Black faculty marginalization through a sequential mixed-methods design. *New Directions for Institutional Research, 2011*(151), 45–61.

Griffin, K. A., Pifer, M. J., Humphrey, J. R., & Hazelwood, A. M. (2011). (Re)defining departure: Exploring Black professors' experiences with and responses to racism and racial climate. *American Journal of Education, 117*(4), 495–526.

Kamimura, A. (2019). *"Untying Our Hands": A mixed methods study of strengths-based approaches to faculty recruitment and hiring practices in STEM for equity* (Doctoral dissertation). Retrieved from ProQuest Dissertations and Theses database. (UMI No. 27614380).

Kayes, P. E. (2006). New paradigms for diversifying faculty and staff in higher education: Uncovering cultural biases in the search and hiring process. *Multicultural Education, 14*(2), 65–69.

Light, P. (1994). "Not like us": Removing the barriers to recruiting minority faculty. *Journal of Policy Analysis and Management*, 164–180.

Linstead, S., & Pullen, A. (2006). Gender as multiplicity: Desire, displacement, difference and dispersion. *Human Relations, 59*(9), 1287–1310.

Merriam, S. B. (2009). *Qualitative research: A guide to design and implementation*. San Francisco, CA: Jossey-Bass.

Moreno, J. F., Smith, D. G., Clayton-Pedersen, A. R., Parker, S., & Teraguchi, D. H. (2006*). The revolving door for underrepresented minorities faculty in higher education: An analysis from the campus diversity initiative*. The James Irving Foundation Campus Diversity Initiative Evaluation Project. Retrieved from www.slcc.edu/inclusivity/docs/the-revolving-door-for-underrepresented-minority-faculty-in-higher-education.pdf

National Science Foundation (2014). *Table 15. Doctorate recipients, by sex and major field: 2003–13*. Retrieved from www.nsf.gov/statistics/sed/2013/data/tab15.pdf

National Science Foundation (2014). *Table 19. Doctorate recipients, by ethnicity, race, and citizenship status: 2003–13*. Retrieved from www.nsf.gov/statistics/sed/2013/data/tab19.pdf

National Science Foundation. (2014). *Table 21. Female doctorate recipients, by ethnicity, race, and citizenship status: 2003–13*. Retrieved from www.nsf.gov/statistics/sed/2013/data/tab21.pdf

National Science Foundation. (2014). *Table 47. Employment sector of doctorate recipients with definite postgraduation U.S. employment commitments, by sex, citizenship status, ethnicity, and race: Selected years, 1993–2013*. Retrieved from www.nsf.gov/statistics/sed/2013/data/tab47.pdf

National Science Foundation, National Center for Science and Engineering Statistics (2013). *Survey of doctorate recipients*. Table 9–22. Arlington, VA: National Center for Science and Engineering Statistics, 2013. Retrieved from www.nsf.gov/statistics/2015/nsf15311/tables/pdf/tab9-22.pdf

National Science Foundation, National Center for Science and Engineering Statistics (2013). *Survey of doctorate recipients*. Table 9–25. Arlington, VA: National Center for Science and Engineering Statistics, 2013. Retrieved from www.nsf.gov/statistics/2015/nsf15311/tables/pdf/tab9-25.pdf

Núñez, A. M., Hurtado, S., & Calderón Galdeano, E. (2015). Why study Hisanic-serving institutions? In A.-M. Nuñez, S. Hurtado, & E. Calderón Galdeano (Eds.), *Hispanic-serving institutions: Advancing research and transformative practice* (pp. 1–24). New York, NY: Routledge.

Olivas, M. (1988). Latino faculty at the border: Increasing numbers key to more Hispanic access. *Change, 20*(3), 6–9.

Reyes, M., & Halcon, J. (1988). Racism in academia: The old wolf revisited. *Harvard Educational Review, 58*(3), 299–315.

Reyes, N. A. S., Carales, V. D., & Sansone, V. A. (2020). Homegrown scholars: A collaborative autoethnography on entering the professoriate, giving back, and coming home. *Journal of Diversity in Higher Education.* http://dx.doi.org/10.1037/dhe0000165

Saenz, V. B. (2010). Breaking the segregation cycle: Examining students' precollege racial environments and college diversity experiences. *The Review of Higher Education, 34*(1), 1–37.

Smith, D. G. (2000). How to diversify the faculty. *Academe, 86*(5), 48–52.

Stanley, C. A. (2006). Coloring the academic landscape: Faculty of color breaking the silence in predominantly White colleges and universities. *American Educational Research Journal, 43*(4), 701–736.

Stewart, A. J., Malley, J. E., & LaVaque-Manty, D. (2010). *Transforming science and engineering: Advancing academic women.* Ann Arbor, MI: University of Michigan Press.

Sutcliffe, K. M., & Vogus, T. J. (2003). Organizing for resilience. In K. S. Cameron, J. E. Dutton, & R. E. Quinn (Eds.), *Positive organizational scholarship.* San Francisco, CA: Berrett-Koehler.

Turner, C. S. V. (2002). *Diversifying the faculty: A guidebook for search committees.* Washington, DC: Association of American Colleges and Universities.

Turner, C. S. V., González, J. C., & Wood, J. L. (2008). Faculty of color in academe: What 20 years of literature tells us. *Journal of Diversity in Higher Education, 1*(3), 139.

Turner, C. S. V., Myers Jr, S. L., & Creswell, J. W. (1999). Exploring underrepresentation: The case of faculty of color in the Midwest. *Journal of Higher Education,* 27–59.

U.S. Department of Education, National Center for Educational Statistics (2008). *Table 315.80. Full-time and part-time faculty and instructional staff in degree-granting postsecondary institutions by race/ethnicity, sex, and program area: Fall 1998 and fall 2003.* Retrieved from https://nces.ed.gov/programs/digest/d18/tables/dt18_315.80.asp

U.S. Department of Education, National Center for Educational Statistics (2011). *National study of postsecondary faculty.* Table 315.20. Washington, DC: U.S. Department of Education. Retrieved from http://nces.ed.gov/programs/digest/d13/tables/dt13_315.20.asp

Yosso, T. J. (2005). Whose culture has capital? A critical race theory discussion of community cultural wealth. *Race Ethnicity and Education, 8*(1), 69–91.

# 5  How Many Latinas in STEM Benefit From High-Impact Practices? Examining Participation by Social Class and Immigrant Status

*Sanga Kim, Selyna Pérez Beverly, and Hyun Kyoung Ro*

Diversity in science, technology, engineering, and mathematics (STEM) fields and in the STEM workforce is an essential national goal, not only for economic development but for gender and racial equity in the United States (National Academy of Sciences, National Academy of Engineering, and Institute of Medicine, 2011). Despite the national-level efforts and investment to improve gender and racial diversity in STEM disciplines, women and racially minoritized groups have been underrepresented in STEM fields for decades, and Latinas are not an exception. In fall 2017, female students accounted for 56% of the total undergraduate enrollment (9.4 million students), and Latina students made up 20.6% of the female undergraduate enrollment (1.9 million students) (National Center for Education Statistics, 2020). Besides the vast underrepresentation of Latinas in universities, their representation in STEM disciplines is also particularly low. Latinas only received 2.31% and 1.87% of bachelor's degrees in engineering and computer science, respectively, in 2016, according to the National Center for Science and Engineering Statistics ([NCSES], 2019). The percentage of Latinas who receive a bachelor's degree in engineering has increased only slightly over time, for example, whereas attainment in computer science has been relatively steady over the past two decades ([NCSES], 2019). While the degree completion of Latinas is higher in biological sciences than other STEM disciplines, Latina students still only account for 7.01% of bachelor's degrees in the field. These national statistics indicate that the goal of diversifying the STEM workforce will not be achieved without increasing the number of Latina engineers and scientists.

This chapter focuses on the potential to increase student engagement to recruit and retain more Latina undergraduates in STEM. Student engagement is defined as students' investment of time and energy into educational activities that universities support and develop (Kuh, Cruce, Shoup, Kinzie, & Gonyea, 2008). Student engagement includes academic experiences in college, enriching educational experiences, supportive learning environments, involvement in co-curricular activities, and interactions with staff, faculty, and peers (Kahu, Stephens, Leach, & Zepke, 2013; Kuh 2009a; Mayhew, Rockenbach, Bowman, Seifert, & Wolniak, 2016). Engagement theory emphasizes that both academic (in-class) engagement and social integration (educationally relevant out-of-class or co-curricular activities) are important to college success (Kuh, 2009a; Tinto, 2006). There is evidence to show that student engagement is positively related to cognitive development, self-esteem, student satisfaction, GPA, and persistence (Kuh, 2009b; Pascarella, Seifert, & Blaich, 2010). Furthermore, student engagement can be particularly beneficial to racially minoritized students (Finley & McNair, 2013; Kuh, 2008). According to Finley and McNair (2013), in their comprehensive report on the effects of high-impact practice (HIP) participation on learning, they found that both Latinx and Black students who participated in HIPs had higher engagement in deep learning and perceived learning gains.

Among racially minoritized students who may receive benefits from HIP participation, we focus on Latina undergraduates in STEM majors at research universities. Researchers have found that for Latinx students, academic and social interaction through curricular and co-curricular activities are positively related to students' sense of belonging (Hurtado & Carter, 1997; York & Fernandez, 2018) and first-year GPA (Kuh, Cruce, Shoup, Kinzie, & Gonyea, et al., 2008). Finley and McNair (2013) found that HIPs can improve both GPA and retention from the first to second year for Latinx students more than White students. The literature suggests that HIP participation can facilitate active engagement with peers, faculty, and staff, and thus can enrich students' educational experiences and benefit student retention and graduation.

Although the relationship between HIPs and outcomes among college students is well documented in higher-education literature (Farinde, Tempest, & Merriweather, 2014; Kilgo & Pascarella, 2016; Koch et al., 2018), the relationship between HIPs and college outcomes of Latinx students focusing on within-group differences remains undeveloped. Scholars also argue for more considerable attention to diversity within the Latinx student population to better describe and document these nuanced experiences that are often overlooked in STEM education research (Cole & Espinoza, 2008; Museus, Palmer, Davis, & Maramba, 2011; Rincón & Lane, 2017; Seymour & Hewitt, 1997; Trent, Owens-Nicholson, & George, 2006; Trusty, 2002). Within the Latinx population, Latinas often experience a "double bind" that refers to unique structural and cultural challenges that underrepresented women of color have faced, as they experience both sexism and racism simultaneously in their STEM careers (Malcom, Hall, & Brown, 1976; Malcom & Malcom, 2011; Ong, Wright, Espinosa, & Orfield, 2011). Connecting to communities and engaging in enriching educational activities through HIP participation would be beneficial to promote and strengthen the academic resilience of Latinas in STEM (Suarez, 2003).

The purpose of this chapter is to better understand the HIP involvement of Latinas pursuing STEM degrees in order to potentially develop ways to increase their persistence and promote their success in college. This chapter also attempts to explore the diversity of Latinas who participate in HIPs by addressing student background characteristics such as social class and immigrant-generation status. We seek to assist practitioners and faculty in identifying ways to encourage Latinas with differing backgrounds to participate in HIPs and ultimately promote their educational opportunities and persistence in STEM.

## Social and Resistant Cultural Capital and Latina STEM Undergraduates

Gender and racial inequities in higher education persist within STEM fields because of the STEM cultural values of meritocracy, which focus on grades and classroom performance, thus dismissing students' social identities (e.g., Carter, Dueñas, Mendoza, 2019; Rainey, Dancy, Mickelson, Stearns, & Moller, 2019). The historical institutionalized practices of science contribute to the idea that science is universal and is dismissive of a person's social identity in scientific practice (Harding, 1998). Harding argues that science excludes anyone who is viewed as "other" to the dominant people, such as women and underrepresented minorities. This exclusion can be amplified for Latinas because they encounter dual systems of oppression in the form of both racism and sexism (Crenshaw, 1991), which can prevent their acceptance not only in college but specifically in STEM spaces, which are dominated by White men.

Studies on Latinx student experiences in college and STEM reveal how social capital and resistant cultural capital contribute to their success and persistence in college (Espinoza, 2013; Martin, Simmons, & Yu, 2013; Sánchez-Connally, 2018). Yosso (2005), in her Community Cultural Wealth Framework, discusses how communities of color build their cultural capital as "an array of knowledge, skills, abilities and contacts possessed and utilized by Communities of Color to survive and resist macro and micro-forms of oppression" (p. 77). In her model, she identifies one of the forms as resistant capital, which is behavior that challenges inequality through resistance.

Resistant capital is evident in Latinx communities as they combat oppression based on race, gender, and immigrant status (Sánchez-Connally, 2018). Sánchez-Connally (2018) studied 21 first-generation Latinx students at four-year college campuses and documented the ways in which they overcame racism, sexism, and classism through resistant cultural capital. Sánchez-Connally (2018) found that Latinx students developed ways to adapt to their environments and succeed. All the participants in the study described their lack of confidence in their academic abilities, which was furthered through negative interactions with peers, faculty, and administrators. In order to overcome the challenges, the students developed a "resistant cultural capital" that relied on counterspaces, such as Latinx affinity groups or organizations, peer networks, and the recognition that they could not quit because of expectations from family as well as their personal expectations.

Peer networks appear to have a significant role in providing both academic assistance and social support for Latinx students, thus contributing to their social capital (Espinoza, 2013; Sánchez-Connally, 2018). Latinx students pursuing STEM degrees appear to benefit from building a community of support made up of friends and family. According to Peralta, Caspary, and Boothe (2013), in their mixed-method study at two universities on Mexican immigrant and first-generation Mexican students, they found that the students were more likely to persist in STEM fields if they had family, friends, and community support. Martin et al. (2013), in a multiple case study of four Latina engineering students in college, found that both peers and institutional supports served as prominent sources of social capital. Therefore, Latinas may benefit from connections to the university, particularly through peers and faculty, in order to persist. The literature suggests that HIPs may be a tool that can foster positive connections for Latinas with faculty, peers, and their community to contribute to their social and resistant cultural capital, particularly if they are navigating exclusionary spaces that consist predominately of White men.

## Effects of High-Impact Practices on Different Populations

HIPs have been identified as a means to further engage students in the university and to achieve specific student learning outcomes (Kilgo, Sheets, & Pascarella, 2015). In HIPs, first-year experience courses, undergraduate research, and both experiential and service learning have proven beneficial to students, especially in STEM (Farinde, Tempest, & Merriweather, 2014; Kilgo & Pascarella, 2016; Koch et al., 2018). Koch et al. (2018) found that first-year experience courses at a Hispanic-Serving Institution contributed to students' "learning power" or their experiences with changing and learning, critical curiosity, meaning-making, creativity, learning relationships, strategic awareness, and resilience (p. 23). Kilgo and Pascarella (2016) found that STEM students' involvement positively influenced student learning, time to degree, and educational goals in faculty research. Farinde et al. (2014) studied college-bound high school students and found that underrepresented students (women, Black, Latinx, and Native American) who participated in a service learning program before college had increased comprehension of engineering concepts, broadened their understanding of the engineering field, and developed an engineering identity by "performing" engineering tasks.

The accessibility and differing developmental needs of minoritized students may be connected to a lack of participation in HIPs (Lange & Stewart,

2019). Lange and Stewart explain that although HIPs vary from institution to institution, they have many aspects in common, which include a focus on writing expression, cognitive and intellectual development, problem solving, and practical application. These particular areas of student learning are valued in academia but frequently can be exclusionary to minoritized populations that may not conform to these dominant forms of epistemology. For instance, traditionally, HIPs place less emphasis on equity, social justice, and community engagement (Lange & Stewart, 2019). Lange and Stewart further suggest that racism and classism may contribute to lower participation of minoritized groups in HIPs, such as low-income students and Students of Color. By considering the implications of social class and immigrant-generation status on Latinas in STEM participation in HIPs and the types of HIPs they participate in, we hope to inform both practitioners and faculty on how to better support Latinas in STEM involvement in HIPs.

## Methods

### Data

This study uses a multi-institutional dataset, the 2016 Student Experience in the Research University (SERU) survey. The 2016 SERU data included 101,280 undergraduates at 18 institutions. Among these students, 39,341 (nearly 41% of the sample) had majors in STEM. Of the students majoring in STEM, 38% were White, 37% Asian, 18% Latinx, 5% Multi-racial, and 2% Black.

### Variables

In this chapter, we examine Latina STEM undergraduates' participation in HIPs, and we examine differences by immigrant background and socioeconomic status. First, we constructed immigrant-generation status in the following way: first-generation immigrant students were classified as those who were born outside the United States and had at least one foreign-born parent; second-generation immigrant students included those who were born in the United States and had at least one foreign-born parent; and non-immigrants were students who were born either in the United States or abroad with two US-native parents (Portes & Rumbaut, 2018). In second-generation immigrant students, we also included those who were born outside the United States and moved there at 5 years of age or younger with at least one foreign-born parent, referred to as the "1.75 generation" because their experiences are, in general, similar to those who were born in the United States in terms of language acquisition and socialization (Rumbaut, 2004). Second, to investigate the link between social class and HIP participation, we categorized socioeconomic status (SES) based on the following question in the SERU survey: "Which of the following best describes your social class when you were growing up: low-income or poor, working-class, middle-class, upper-middle or professional-middle, and wealthy." We first combined "low-income or poor" and "working-class" to create a "low-SES" group. We also combined "upper-middle or professional-middle" and "wealthy" to create a "high-SES" group.

Finally, we examined 13 specific HIPs, which students either completed or were in the process of completing when they responded to the 2016 SERU survey. The HIPs included (a) first-year seminar; (b) capstone or thesis projects; (c) writing-intensive/enriched courses; (d) learning community, taking two or more linked classes with the same cohort of students; (e) living-learning program(s), where students with common interests live together and share learning experiences in and out of the classroom; (f) leadership program; (g) honors program; (h) undergraduate research, research or creative project outside of regular course requirements; (i) internship, either credit or non-credit bearing; (j) on-campus academic experiences with an international/global focus; (k) study abroad, academically-focused time outside of the United States in which at least one academic credit is accrued; (l) academic service learning or community-based learning experience (service learning); and (m) academic experiences with diversity (e.g., race, gender, sexual orientation) focus.

### Sample

The final analytical sample included 3,453 Latina undergraduate students in STEM disciplines at 18 research universities. Within this sample, 64% identified as low-SES, 27% were middle-SES, and 9% were high-SES. In terms of immigrant background, 8% were first-generation immigrants, 66% were second-generation immigrants, and 26% were non-immigrant students. In this chapter, the descriptive data indicates the percentage (out of 100%) of participation in HIPs within each group of students (e.g., first-generation immigrant Latina in STEM, low-SES Latina in STEM).

### Findings

In order to present our findings on the 13 HIPs in a cohesive manner, we grouped the HIPs according to similar attributes. We first utilized the National Survey of Student Engagement (NSSE) (2020) broad descriptions of the six primary high-impact practices they ask students about on their annual survey. These HIPs include (1) learning community or another formal program where students take two or more courses together; (2) course with a community-based project such as service learning; (3) engaging in research with a faculty member; (4) internship, co-op, field experience, student teaching, or clinical placement; (4) study abroad; and (5) culminating senior experience which includes capstone, senior project, comprehensive exam, portfolio, etc. We used these broad categories to inform how we would group the different HIPs in our sample.

We first combined first-year seminar, capstone or thesis project, and writing-intensive courses into one category. Although the first-year seminar is typically taken in the first year and writing-intensive courses can be taken throughout college, we decided to combine these categories with the capstone or thesis project because they share a curricular academic component. Second, we combined learning communities, living-learning communities, leadership programs, and honors programs, given that they potentially involve students taking classes and engaging in outside classroom activities together as a group. Third, we combined undergraduate research and internships because of their apprentice-type experiences in which they can potentially assist students in identifying graduate education and career interests. Fourth, global-focused experiences and study abroad were placed together because they involve assisting students in developing cultural competency. Lastly, service learning and academic experiences with diversity were made into one category because they are typically associated with a course and focus on exposure to diverse communities.

### First-Year Seminar, Capstone Projects, and Writing-Intensive Courses

Figure 5.1 shows that approximately one-tenth of STEM Latinas participated in capstone or thesis projects, and approximately half of STEM Latinas participated in first-year seminar and writing-intensive courses. It is worth noting that more than 90% of higher-education institutions offer some type of a first-year seminar, but there are considerable variations in terms of who takes the course (e.g., whether it is required for all first-year students, voluntary, mandatory for students who meet specific requirements or at-risk students) (Clark & Cundiff, 2011; Tobolowsky, Mamrick, & Cox, 2005).

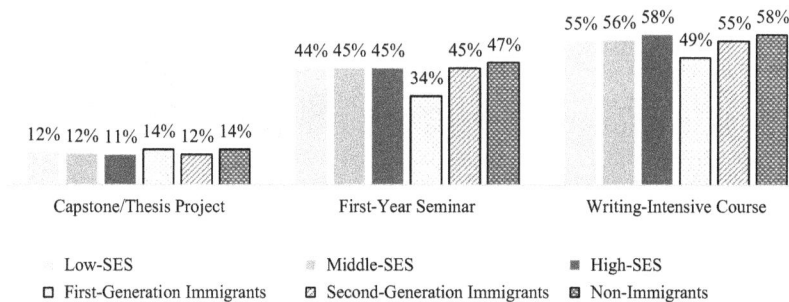

*Figure 5.1* Capstone, First-Year Seminar, and Writing-Intensive Course Participation by Latina Socioeconomic and Immigrant-Generation Status.

Note: Figure 5.1 is based on the authors' data analysis from the 2016 SERU survey.

The 2016 SERU data indicates that almost half of STEM Latina students participate in a first-year seminar, except for the first-generation immigrant group.

While we found similar patterns in capstone participation by socioeconomic status and immigrant status among STEM Latinas, we found that first-generation immigrants participated at a lower rate than their peers in first-year seminar and writing-intensive courses. When we compared the percentage of participation in these practices within each group of students, the second-generation and non-immigrant STEM Latina students were 11 percentage points (45%) and 13 percentage points (47%) higher in first-year seminar participation than were first-generation immigrants (34%), respectively. The participation of first-generation immigrant STEM Latinas in writing-intensive courses (49%) was 6 percentage points lower than second-generation immigrants (55%) and 9 percentage points lower than non-immigrants (58%). The gap between second-generation immigrants and non-immigrants in these HIPs was much smaller than the comparison with first-generation immigrants among STEM Latinas. This finding suggested that the distribution of participation in first-year seminar and writing-intensive courses differs among STEM Latinas with varying immigrant backgrounds.

### Learning Communities, Living-Learning Communities, Leadership, and Honors Programs

Figure 5.2 indicates that low-SES (43%) and middle-SES students (44%) participated in learning communities at levels that were almost ten percentage points higher than the high-SES STEM Latinas (35%). We do not know whether universities intend to recruit low- and middle-SES students for learning communities or whether these students are more interested in

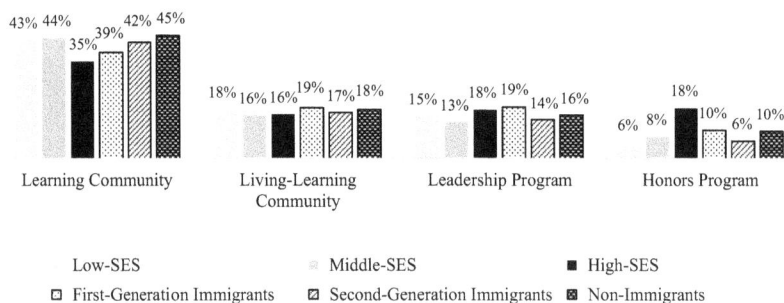

*Figure 5.2* Learning Community, Living-Learning Community, Leadership Program, and Honors Program Participation by Latina Socioeconomic and Immigrant-Generation Status.

Note: Figure 5.2 is based on the authors' data analysis from the 2016 SERU survey.

participating in learning communities, compared to high-SES peers. Either way, learning communities would be beneficial for economically disadvantaged students. Tinto (2019) argued that learning communities can be beneficial to historically underrepresented or economically disadvantaged students because they learn skills or strategies to overcome social or academic challenges. Research shows that low-SES and middle-SES students may get more benefit from learning communities than high-SES students by supporting students' adjustment to college (Cambridge-Williams, Winsler, Kitsantas, & Bernard, 2013) and collaborative learning activities and interactions with peers of different backgrounds (Tinto, 2019).

While we found higher proportions of low-SES and middle-SES STEM Latinas in learning and living-learning communities, we found a reverse pattern in honors programs. Most higher-education institutions have criteria to be eligible for participation in honors programs; students need to demonstrate their academic ability and achievement (e.g., high school grade point average and ACT/SAT scores) (Achterberg, 2005; Bowman & Culver, 2018). Given the social-class achievement gap, low-SES students who, in general, have lower academic performance may not be able to access honors programs, or may not even be aware of honors programs. The finding of participation in honors programs is consistent with prior literature; low-SES (6%) and middle-SES (8%) groups were ten percentage points lower than high-SES students (18%). The percent of low-SES Latinas participating in honors programs (6%) is the lowest among all 13 HIPs. Given that researchers have found positive effects of honors programs on student learning outcomes (Bowman & Culver, 2018; Seifert, Pascarella, Colangelo, & Assouline, 2007), we recommend that administrators consider broadening honors programs to low-SES STEM Latina students by offering tutoring and faculty mentoring. Low-SES Latina students should not be excluded because of their academic preparedness, which is tied to broader contexts like parental education and expectations, school location and resources, and distribution of wealth (Oakes, 2003).

We did not find substantial differences in living-learning community and leadership program participation by socioeconomic and immigrant-generation status. These trends may reveal that Latinas with low SES may already be involved in STEM intervention programs or other social identity groups based on race/ethnicity that function as living-learning communities or leadership programs. Analyzing learning outcome differences by gender and race among 5,000 engineering undergraduates at 31 institutions, Ro and Loya (2015) found that Latina undergraduates in engineering programs self-assessed their leadership skills higher than their White male peers. There are no differences in any learning outcomes that Ro and Loya (2015) measured (e.g., fundamental, design, communication, teamwork, and contextual competence) between Latinas and White men. Leadership skills were the only case in which a minoritized female group (Blacks, Latinas, and Asian Americans) surpassed White men's self-reports. STEM intervention programs that target underrepresented racial minorities as well as social identity groups, such as the Society of Hispanic Professional Engineers, focus on building a sense of community and facilitating the development of a science identity and leadership skills (Maton et al., 2016). STEM Latinas may boost their confidence in leadership through leadership programs as well as STEM intervention programs.

### Undergraduate Research and Internship

Figure 5.3 indicates noticeable differences by students' socioeconomic status in participation in undergraduate research and internships. We found that high-SES Latinas completed these two practices at almost 10 to 15 percentage points higher than either middle-SES or low-SES Latinas. This finding is consistent with previous literature that indicates low-SES students have fewer opportunities to be involved in undergraduate research or internships because of a lack of parental social and cultural capital, academic preparedness, and interaction with faculty (Boylan, 2009; Carter, Ro, Alcott, & Lattuca, 2016). Extending undergraduate research opportunities and internships to low-SES students is critical because these apprentice experiences, particularly in STEM, can promote student learning outcomes (Fernald & Goldstein, 2013; Jones, 2002; Kilgo & Pascarella, 2016; Parker III, Kilgo, Sheets, & Pascarella, 2016).

Across students' immigrant statuses, we found a higher proportion of first-generation immigrants in undergraduate research (40%) compared to either second-generation (28%) or non-immigrant Latina students (32%). We found a similar pattern in internship participation: first-generation immigrants (48%);

Low-SES        Middle-SES
■ High-SES      □ First-Generation Immigrants
☒ Second-Generation Immigrants    ⊠ Non-Immigrants

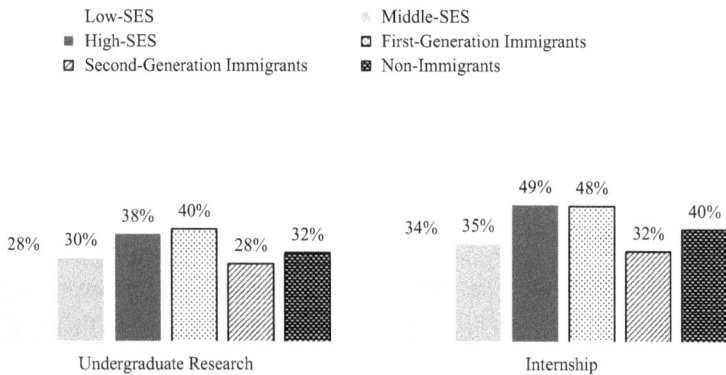

*Figure 5.3* Undergraduate Latina Research and Internship Participation by Socioeconomic and Immigrant-Generation Status.

Note: Figure 5.3 is based on the authors' data analysis from the 2016 SERU survey.

second-generation immigrants (32%); and non-immigrants (40%). Given that the SERU did not ask more specific questions about HIPs, it is uncertain what kinds of undergraduate research and internship programs first-generation immigrant STEM Latinas tend to participate in while in college. This finding may be connected to involvement in STEM intervention programs or other university transition programs that focus efforts on recruiting and retaining underrepresented students, which include those who are first-generation immigrants.

### Global-Focused Academic Experiences and Study Abroad

Figure 5.4 shows that there are no substantial differences in global-focused experiences and study abroad across SES and immigration status among STEM Latinas. Study abroad programs can be cost-prohibitive, which may be a deterrent for low-SES students. However, accessibility and cultural aspects may contribute to decisions of a small percentage of both economically disadvantaged and advantaged STEM Latinas to choose to participate in global-focused experiences and study abroad. Additionally, both first- and second-immigrant and non-immigrant students may want or need to be close to home and their families, which may also discourage participation in study abroad programs.

### Service Learning and Academic Experiences With Diversity

Figure 5.5 shows that there is no substantial difference in learning and academic experiences with diversity by social class and immigrant status. While about one-third of STEM Latinas participated in service learning, approximately 51–58% of STEM Latinas participated in academic experiences with diversity, regardless of their social class and immigrant background.

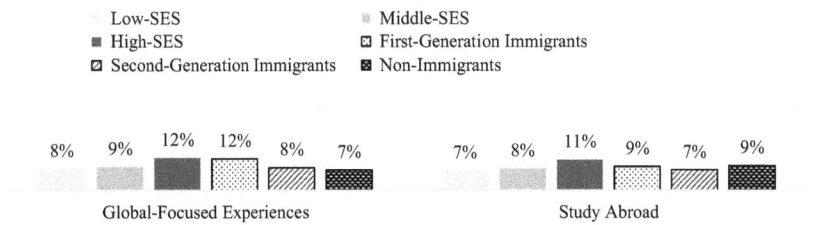

Legend:
- Low-SES
- High-SES
- Second-Generation Immigrants
- Middle-SES
- First-Generation Immigrants
- Non-Immigrants

Global-Focused Experiences: 8% 9% 12% 12% 8% 7%

Study Abroad: 7% 8% 11% 9% 7% 9%

*Figure 5.4* Global-Focused Academic Latina Experiences and Study Abroad Participation by Socioeconomic and Immigrant-Generation Status.

Note: Figure 5.4 is based on the authors' data analysis from the 2016 SERU survey.

Low-SES          Middle-SES          ■ High-SES
First-Generation Immigrants    ☑ Second-Generation Immigrants  ▨ Non-Immigrants

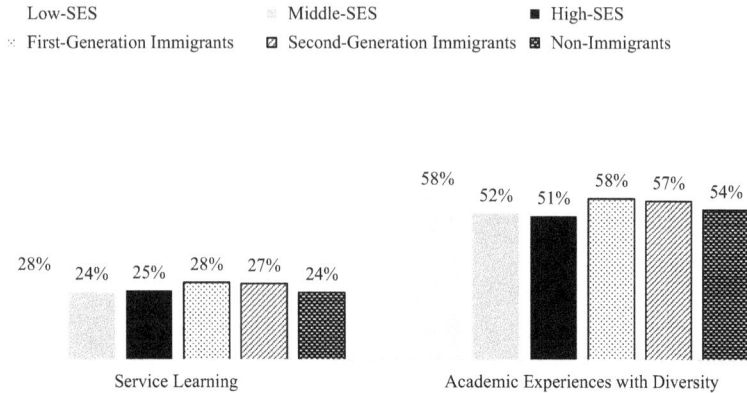

*Figure 5.5* Service Learning and Academic Latina Experiences With Diversity Participation by Socioeconomic and Immigrant-Generation Status.

Note: Figure 5.5 is based on the authors' data analysis from the 2016 SERU survey.

## Summary and Recommendations

### Conclusions

This chapter examines differences in the types of HIPs that Latinas in STEM completed based on their socio-demographic characteristics. Across thirteen HIPs examined in this chapter, nearly half of the entire analytical sample of Latinas in STEM participated in first-year seminar, learning communities, writing-intensive courses, and academic experiences with diversity. This finding provides insight into the type of HIPs that STEM Latinas at research universities are engaging in. These experiential activities may provide Latinas with the opportunity to increase interaction with peers, faculty, and staff, which are positively related to student learning outcomes and persistence (Kuh, et al., 2005; Zhao, Carini, & Kuh, 2005). Also, these connections may contribute to Latinas' resistant cultural capital and social capital as they develop communities that assist them with combating both sexism and racism at the university. Particularly in STEM disciplines, writing-intensive courses can promote critical thinking, problem-solving abilities, (Dowd, Thompson, Schiff, & Reynolds, 2018; Hodges, 2015) and scientific reasoning (Reynolds, Thaiss, Katkin, & Thompson, 2012).

Our findings also indicate that less than 10% of the Latinas in STEM in our sample completed honors programs, international/global-focused academic experiences, and study abroad programs. A key element in international/global-focused academic experiences and study abroad programs is the opportunity for students to develop cultural competency and explore different worldviews from their own (Kuh, 2008). In higher education, a study abroad experience is positively related to growth in sociocultural awareness, cultural competence, (Bell, Gibson, Tarrant, Perry, & Stoner, 2016; Kilgo, Sheets, & Pascarella, 2015) and creative thinking (Lee, Therriault, & Linderholm, 2012). Compared to the HIPs that showed a high participation rate of Latinas in STEM, these programs seek to provide students with diverse and broad perspectives beyond campus and a wide range of interactions in the community. This finding suggests that we need a better understanding of why Latinas in STEM choose certain HIPs over others, as well as the institutional and departmental structures that may prevent participation of Latinas in STEM in particular HIPs.

We also found variation across Latinas' self-identified socioeconomic statuses and immigrant backgrounds. For low-SES Latinas in STEM, their participation in honors programs is 12 percentage points lower than high-SES Latinas in STEM. Low-SES Latina students also indicate a lower percentage of participation in undergraduate research and internship programs compared to their middle-SES or high-SES counterparts. STEM Latina undergraduates, especially those who are economically disadvantaged or have immigrant backgrounds, may lack the resources to navigate college and take advantage of programs that enhance academic, professional, and socio-emotional growth (Martin, Simmons, & Yu, 2013). These differences in HIP participation warrant attention from policymakers and practitioners.

### Recommendations for Practice and Future Research

First, to maximize the positive impact of HIPs for Latinas in STEM—and in particular, those with low-SES and immigrant backgrounds—faculty and staff may need to find ways to provide structured opportunities and customized practices to meet the unique needs and expectations of different student subgroups (Kuh et al., 2005; Kuh, Cruce, Shoup, Kinzie, & Gonyea, 2008). For instance, STEM intervention programs that intentionally focus on developing a supportive network and academic support may seek to increase the number of Latinas with a low social class background. Second, faculty and staff should examine how information-sharing pipelines have been structured and implemented to prompt equal educational opportunities and broaden HIP participation. Zhao, Carini, and Kuh (2005) found that women in STEM majors were engaged in the more formal and public forms of effective educational practices (e.g., study abroad, internships, and senior capstone courses) than their male counterparts; however, they less frequently collaborated on projects with peers and less often discussed the ideas from class or readings with faculty outside of class. Even though both men and women participate in HIPs, STEM undergraduate women may have limited access to information exchanges and informal learning opportunities with their peers and faculty. Faculty and administrators need to find ways to establish infrastructure to improve equal educational opportunities while implementing HIPs, especially for Latinas who are also low-SES. When institutions design the programs and curricula of HIPs, they should reflect upon the backgrounds of Latinas with diverse backgrounds in STEM.

This finding also lays the foundation for future research that can begin to analyze to what extent the involvement of Latinas in STEM in HIPs affects certain learning outcomes, and their sense of belonging. We also need to explore the differential effects of HIP participation by student background characteristics within the Latinx population. While there is established evidence suggesting the positive effects of HIPs on student learning and college outcomes for all students, limited research has examined the conditional effects of HIPs based on Latina students' background characteristics (e.g., Bai & Pan, 2009; Bowman & Culver, 2018; Seifert, Gilling, Hanson, Pascarella, & Blaich, 2014; Seifert, Pascarella, Colangelo, & Assouline, 2007). For researchers, the descriptive data presented throughout this chapter reveal differences among Latinas in social class and immigrant-generation status and suggest that their college experiences and academic and social engagement may interact with their background characteristics. More research is needed to understand the impact of HIPs on Latinas' socio-demographic characteristics (e.g., parental level of education, pre-college preparedness, sexual orientation, or transfer status) to successfully estimate the impact of HIPs on student learning outcomes (Seifert, Gilling, Hanson, Pascarella, & Blaich, 2014).

Further research is also needed to investigate the impact of HIPs on Latinas majoring in sub-STEM disciplines separately. Particularly, STEM disciplines in which women and Students of Color are historically underrepresented (e.g., computer science, engineering, physics, mathematics, and statistics) should receive attention. Compared with biological sciences, peer relationships with students who share the same gender play a significant role in increasing women students' sense of belonging in physical science and engineering disciplines (e.g., Dasgupta, 2011). Sense of belonging seems to be more beneficial for women students' academic motivation and persistence intentions in these STEM disciplines than for men (Lewis et al., 2017). Therefore, HIP participation may contribute to students' perceptions of departmental/institutional climate, sense of belonging, and learning outcomes, which may differ between women and men in STEM disciplines. Overall, continued efforts to understand variations among Latinas and their relationship with HIPs will help improve institutional policies and practices to support the persistence of Latinas in STEM.

# References

Achterberg, C. (2005). What is an honors student? *Journal of the National Collegiate Honors Council*, 6(1), 75–83.

Bai, H., & Pan, W. (2009). A multilevel approach to assessing the interaction effects on college student retention. *Journal of College Student Retention: Research, Theory & Practice*, 11(2), 287–301.

Bell, H. L., Gibson, H. J., Tarrant, M. A., Perry, L. G., III, & Stoner, L. (2016). Transformational learning through study abroad: US students' reflections on learning about sustainability in the South Pacific. *Leisure Studies*, 35(4), 389–405.

Bowman, N. A., & Culver, K. C. (2018). When do honors programs make the grade? Conditional effects on college satisfaction, achievement, retention, and graduation. *Research in Higher Education*, 59(3), 249–272.

Boylan, M. (2009). Undergraduate STEM research experiences: Impact on student interest in doing graduate work in STEM fields. In R. G. Ehrenberg & C. V. Kuh (Eds.), *Doctoral education and the faculty of the future* (pp. 109–120). Ithaca, NY: Cornell University Press.

Cambridge-Williams, T., Winsler, A., Kitsantas, A., & Bernard, E. (2013). University 100 orientation courses and living-learning communities boost academic retention and graduation via enhanced self-efficacy and self-regulated learning. *Journal of College Student Retention: Research, Theory & Practice*, 15(2), 243–268.

Carter, D. F., Dueñas, J. E. R., & Mendoza, R. (2019). Critical examination of the role of STEM in propagating and maintaining race and gender disparities. In M. B. Paulsen & L. W. Perna (Eds.), *Higher education: Handbook of theory and research* (Vol. 34, pp. 39–97). Cham, Switzerland: Springer.

Carter, D. F., Ro, H. K., Alcott, B., & Lattuca, L. R. (2016). Co-curricular connections: The role of undergraduate research experiences in promoting engineering students' communication, teamwork, and leadership skills. *Research in Higher Education*, 57(3), 363–393.

Clark, M. H., & Cundiff, N. L. (2011). Assessing the effectiveness of a college freshman seminar using propensity score adjustments. *Research in Higher Education*, 52(6), 616–639.

Cole, D., & Espinoza, A. (2008). Examining the academic success of Latino students in science technology engineering and mathematics (STEM) majors. *Journal of College Student Development*, 49(4), 285–300.

Crenshaw, K. (1991). Mapping the margins: Intersectionality, identity politics, and violence against women of color. *Stanford Law Review*, 43(6), 1241–1299.

Dasgupta, N. (2011). Ingroup experts and peers as social vaccines who inoculate the self-concept: The stereotype inoculation model. *Psychological Inquiry*, 22(4), 231–246.

Dowd, J. E., Thompson Jr, R. J., Schiff, L. A., & Reynolds, J. A. (2018). Understanding the complex relationship between critical thinking and science reasoning among undergraduate thesis writers. *CBE—Life Sciences Education*, 17(1), article 4.

Espinoza, A. (2013). The college experiences of first-generation college Latino students in engineering. *Journal of Latino/Latin America Studies*, 5(2), 71–84.

Farinde, A. A., Tempest, B., & Merriweather, L. (2014). Service learning a bridge to engineering for underrepresented minorities. *International Journal for Service Learning in Engineering*, 475–491.

Fernald, P. S., & Goldstein, G. S. (2013). Advanced internship: A high-impact, low-cost, super-capstone course. *College Teaching*, 61(1), 3–10.

Finley, A., & McNair, T. (2013). *Assessing underserved students' engagement in high-impact practices.* Washington, DC: Association of American Colleges and Universities.

Harding, S. (1998). *Is science multicultural? Postcolonialism, feminism and epistemologies.* Indianapolis, IN: Indiana University Press.

Hodges, L. C. (2015). *Teaching undergraduate science: A guide to overcoming obstacles to student learning.* Sterling, VA: Stylus Publishing, LLC.

Hurtado, S., & Carter, D. F. (1997). Effects of college transition and perceptions of the campus racial climate on Latino college students' sense of belonging. *Sociology of Education*, 324–345.

Jones, E. A. (2002). *Transforming the curriculum: Preparing students for a changing world.* ASHE-ERIC Higher Education Report. Jossey-Bass Higher and Adult Education Series. Jossey-Bass, 989 Market Street, San Francisco, CA 94103–1741.

Kahu, E. R., Stephens, C., Leach, L., & Zepke, N. (2013). The engagement of mature distance students. *Higher Education Research & Development*, 32(5), 791–804.

Kilgo, C. A., & Pascarella, E. T. (2016). Does independent research with a faculty member enhance four-year graduation and graduate/professional degree plans? Convergent results with different analytical methods. *Higher Education*, 71(4), 575–592.

Kilgo, C. A., Sheets, J. K. E., & Pascarella, E. T. (2015). The link between high-impact practices and student learning: Some longitudinal evidence. *Higher Education*, 69(4), 509–525.

Koch, R., Kucsera, J., Angus, K. B., Norman, K., Bowers, E., Nair, P., Moon, H. S., Karimi, A., & Barua, S. (2018). Enhancing learning power through first year experiences for students majoring in STEM disciplines. *Journal of STEM Education*, 19(1), 22–30.

Kuh, G. D. (2008). *High-impact educational practices: What they are, who has access to them, and why they matter.* Washington, DC: Association of American Colleges and Universities.

Kuh, G. D. (2009a). The national survey of student engagement: Conceptual and empirical foundations. *New Directions for Institutional Research*, 2009(141), 5–20.

Kuh, G. D. (2009b). What student affairs professionals need to know about student engagement. *Journal of College Student Development*, 50(6), 683–706.

Kuh, G. D., Cruce, T. M., Shoup, R., Kinzie, J., & Gonyea, R. M. (2008). Unmasking the effects of student engagement on first-year college grades and persistence. *The Journal of Higher Education*, 79(5), 540–563.

Kuh, G. D., Kinzie, J., Schuh, J. H., Whitt, E. J., & Associates (2005). *Student success in college: Creating conditions that matter.* San Francisco, CA: Jossey Bass.

Lange, A. C., & Stewart, D. L. (2019). High-impact practices. In E. S. Abes, S. R. Jones, & D. L. Stewart (Eds.), *Rethinking college student development theory using critical frameworks.* Sterling, VA: Stylus.

Lee, C. S., Therriault, D. J., & Linderholm, T. (2012). On the cognitive benefits of cultural experience: Exploring the relationship between studying abroad and creative thinking. *Applied Cognitive Psychology*, 26(5), 768–778.

Lewis, K. L., Stout, J. G., Finkelstein, N. D., Pollock, S. J., Miyake, A., Cohen, G. L., & Ito, T. A. (2017). Fitting in to move forward: Belonging, gender, and persistence in the physical sciences, technology, engineering, and mathematics (pSTEM). *Psychology of Women Quarterly*, 41(4), 420–436.

Malcom, L. E., & Malcom, S. M. (2011). The double bind: The next generation. *Harvard Educational Review*, 81(2), 162–172.

Malcom, S. M., Hall, P. Q., & Brown, J. W. (1976). *The double bind: The price of being a minority woman in science* (No. 76-R-3). Washington, DC: American Association for the Advancement of Science.

Martin, J. P., Simmons, D. R., & Yu, S. L. (2013). The role of social capital in the experiences of Hispanic women engineering majors. *Journal of Engineering Education, 102*(2), 227–243.

Maton, K. I., Beason, T. S., Godsay, S., Sto. Domingo, M. R., Bailey, T. C., Sun, S., & Hrabowski III, F. A. (2016). Outcomes and processes in the Meyerhoff scholars program: STEM PhD completion, sense of community, perceived program benefit, science identity, and research self-efficacy. *CBE—Life Sciences Education, 15*(3), article 48.

Mayhew, M. J., Rockenbach, A. N., Bowman, N. A., Seifert, T. A., & Wolniak, G. C. (2016). *How college affects students: 21st century evidence that higher education works*. San Francisco, CA: John Wiley & Sons.

Museus, S. D., Palmer, R. T., Davis, R. J., & Maramba, D. C. (2011). *Racial and ethnic minority students' success in STEM education*. Jossey-Bass Incorporated.

National Academy of Sciences, National Academy of Engineering, and Institute of Medicine (2011). *Expanding underrepresented minority participation: America's science and technology talent at the crossroads*. Washington, DC: The National Academies Press. https://doi.org/10.17226/12984

National Center for Education Statistics (2020). Table 306.10. Total fall enrollment in degree granting postsecondary institutions, by level of enrollment, sex, attendance status, and race/ethnicity or nonresident alien status of student: Selected years, 1976 through 2017. Retrieved February 20, 2020, from https://nces.ed.gov/programs/digest/d18/tables/dt18_306.10.asp

National Center for Science and Engineering Statistics [NCSES] (2019). *Women, minorities, and persons with disabilities in science and engineering: 2019 (NSF 19–304)*. Retrieved from www.nsf.gov/statistics/wmpd

National Survey of Student Engagement (2020). Retrieved from http://nsse.indiana.edu/

Oakes, J. (2003). *Critical conditions for equity and diversity in college access: Informing policy and monitoring results*. Los Angeles: University of California All Campus Consortium on Research for Diversity, UC Berkeley.

Ong, M., Wright, C., Espinosa, L., & Orfield, G. (2011). Inside the double bind: A synthesis of empirical research on undergraduate and graduate women of color in science, technology, engineering, and mathematics. *Harvard Educational Review, 81*(2), 172–209.

Parker III, E. T., Kilgo, C. A., Sheets, J. K. E., & Pascarella, E. T. (2016). The differential effects of internship participation on end-of-fourth-year GPA by demographic and institutional characteristics. *Journal of College Student Development, 57*(1), 104–109.

Pascarella, E. T., Seifert, T. A., & Blaich, C. (2010). How effective are the NSSE benchmarks in predicting important educational outcomes? *Change: The Magazine of Higher Learning, 42*(1), 16–22.

Peralta, C., Caspary, M., & Boothe, D. (2013). Success factors impacting Latina/o persistence in higher education leading to STEM opportunities. *Cultural Studies of Science Education, 8*(4), 905–918.

Portes, A., & Rumbaut, R. G. (2018). *Children of Immigrants Longitudinal Study (CILS), San Diego, California, Ft. Lauderdale and Miami, Florida, 1991–2006*. Ann Arbor, MI:

Inter-university Consortium for Political and Social Research [distributor]. Retrieved from https://doi.org/10.3886/ICPSR20520.v

Rainey, K., Dancy, M., Mickelson, R., Stearns, E., & Moller, S. (2019). A descriptive study of race and gender differences in how instructional style and perceived professor care influence decisions to major in STEM. *International Journal of STEM Education*, *6*(1), 1–13.

Reynolds, J. A., Thaiss, C., Katkin, W., & Thompson Jr, R. J. (2012). Writing-to-learn in undergraduate science education: A community-based, conceptually driven approach. *CBE—Life Sciences Education*, *11*(1), 17–25.

Rincón, B. E., & Lane, T. B. (2017). Latin@ s in science, technology, engineering, and mathematics (STEM) at the intersections. *Equity & Excellence in Education*, *50*(2), 182–195.

Ro, H. K., & Loya, K. I. (2015). The effect of gender and race intersectionality on student learning outcomes in engineering. *The Review of Higher Education*, *38*(3), 359–396.

Rumbaut, R. G. (2004). Ages, life stages, and generational cohorts: Decomposing the immigrant first and second generations in the United States. *International Migration Review*, *38*(3), 1160–1205.

Sánchez-Connally, P. (2018). Latinx first generation college students: Negotiating race, gender, class, and belonging. *Race, Gender & Class*, 25.

Seifert, T. A., Gillig, B., Hanson, J. M., Pascarella, E. T., & Blaich, C. F. (2014). The conditional nature of high impact/good practices on student learning outcomes. *The Journal of Higher Education*, *85*(4), 531–564.

Seifert, T. A., Pascarella, E. T., Colangelo, N., & Assouline, S. G. (2007). The effects of honors program participation on experiences of good practices and learning outcomes. *Journal of College Student Development*, *48*(1), 57–74.

Seymour, E., & Hewitt, N. M. (1997). *Talking about leaving: Why undergraduates leave the sciences*. Boulder, CO: Westview.

Suarez, A. L. (2003). Forward transfer: Strengthening the educational pipeline for Latino community college students. *Community College Journal of Research &Practice*, *27*(2), 95–117.

Tinto, V. (2006). Research and practice of student retention: What next? *Journal of College Student Retention: Research, Theory and Practice*, *8*(1), 1–19.

Tinto, V. (2019). Learning better together. *Transitioning Students in Higher Education: Philosophy, Pedagogy and Practice*, 2.

Tobolowsky, B. F., Mamrick, M., & Cox, B. E. (2005). *The 2003 national survey of first-year seminars: Continuing innovations in the collegiate curriculum* (Monograph no. 41). Columbia, SC: National Resource Center for the First Year Experience & Students in Transition, University South Carolina.

Trent, W. T., Owens-Nicholson, D., & George, C. E. (2006). *GMS effect on diversifying math, science, computer science and engineering*. A report to the Bill and Melinda Gates Foundation.

Trusty, J. (2002). Effects of high school course-taking and other variables on choice of science and mathematics college majors. *Journal of Counseling & Development*, *80*, 464.

York, T. T., & Fernandez, F. (2018). The positive effects of service-learning on transfer students' sense of belonging: A multi-institutional analysis. *Journal of College Student Development*, *59*(5), 579–597.

Yosso, T. J. (2005). Whose culture has capital? A critical race theory discussion of community cultural wealth. *Race Ethnicity and Education, 8*(1), 69–91.

Zhao, C. M., Carini, R. M., & Kuh, G. D. (2005). Searching for the peach blossom Shangri-La: Student engagement of men and women SMET majors. *The Review of Higher Education, 28*(4), 503–525.

# Part 2

# Reading (Hearing) Testimonios of Latinas in STEM

# 6 Empowering Latina STEM Majors at a Public R1 Doctoral Hispanic-Serving Institution in Texas

Strategies for Success

*Elsa M. Gonzalez, Mauricio Molina, and Sarah Churchill Turner*

## Introduction

Student environment within science, technology, engineering, and mathematics (STEM) disciplines has a profound impact on females' professional and academic experiences. Those experiences differ greatly for females in comparison to their male peers because of gender representation differences. For Latinas, these experiential differences are influenced by ethnic representation as well. However, despite Latinas being both female and Students of Color—both of which are historically underrepresented groups in STEM (Rainey, Dancy, Mickelson, Stearns, & Moller, 2018)—trends have been improving. These advances become evident when the enrollment trends for Latinas going into higher education are considered. Nationally, from 2000 to 2015, college enrollment for Hispanics increased from 22% to 37%, while Latinas saw an even greater enrollment rate increase—from 25% to 41% (McFarland et al., 2017). These statistics place Hispanic students behind only White and Asian students in undergraduate enrollment at four-year institutions.

Although Hispanics are the fastest-growing minority group in higher education within the past 20 years, with a 125% population increase from 2000 to 2015 (Fleming & Mcphail, 2019), low representation within STEM programs continues. Of all STEM bachelor's degrees conferred in 2014–2015, only 10% went to Hispanic graduates, while 66% were conferred to White graduates (McFarland et al., 2017). At the master's level, the amount of STEM degrees conferred to Hispanic students was only 8%, while White graduates received 65% (McFarland et al., 2017). If only Latinas are considered in STEM, these ratios decrease even further. For all STEM degrees earned in 2011–2012, Latinas were decidedly inadequately represented. In fact, this uneven representation in STEM between males and females is evident across ethnicity, race, and citizenship. Consequently, the need for greater representation of Hispanic students, especially Latina students, pursuing degrees in and working within the STEM fields is higher than ever before (Ong, Wright, Espinosa, & Orfield, 2011).

Thus, researchers in the study detailed in this chapter sought to explore and understand the success stories of Latina students who persist in STEM fields at a public Research 1 (R1) Doctoral university, also designated as a Hispanic-serving institution (HSI), in Texas. The research objective of the proposed study was to identify specific factors, including culture, background, and family support, that both contribute to resilience and create barriers that compromise resilience in Latina STEM majors. This knowledge will help inform evidence-based interventions, strategies, and policies to enhance the success of Latinas—and perhaps other underrepresented minority college students—in STEM. Latina higher-education stories of success are critically needed in the research literature. Therefore, it is important to gain insights from Latina STEM majors to improve strategies to recruit, retain, and graduate these students. To do so, this study was guided by the following three research questions:

1. How does resilience in Latina students—in relation to culture, background, and family support—play a role in their persisting toward the completion of a college STEM degree?
2. How does social climate influence Latina students in the STEM fields?
3. What types of successful strategies do Latinas develop in their pursuit of a college STEM degree?

## Conceptual Framework

This study was guided by resilience theories. Stewart, Reid, and Mangham (1997) implied that resilience can be considered an intricate, dynamic, and biopsychosocial/spiritual process that depends on life context. This overarching concept applies to individuals, families, and communities (Van Breda, 2001). Resilience is also influenced by diversity, which includes ethnicity, race, gender, and more. At the individual level, the cultural identity component can influence resilience, especially within oppressed groups (Van Breda, 2001). When individuals face difficult adversity, resilience is usually viewed as a protective, beneficial asset that, though often dormant, emerges from within an individual in times of struggle.

Resilience involves internal and external factors that support the ability to manage and conquer the most demanding obstacles. Internal factors include attitudes, while external factors include community well-being (Greene, Galambos, & Lee, 2004). In addition, the following characteristics are often discussed when examining the development of resilience: faith, spirituality, and the belief in something greater than oneself. Caring relationships (Ainsworth, 1989) and strength in support communities (Steele & Steele, 1994) are additional areas that research has demonstrated contribute to resilience (Greene et al., 2004). Multiple factors, such as parental communication and one's attitude toward mathematics (or other STEM-related disciplines), can also work together to nurture a sense of resilience within Latinas (Boutin-Martinez, Mireles-Rios, Nylund-Gibson, & Simon, 2019). Such specific factors, which manifest at the high-school level, can influence Latinas' persistence into the postsecondary level, specifically in STEM. Understanding the various factors that can play a role in resilience building is essential in understanding how Latinas persist in STEM.

These aspects of resilience theories, when applied to Latinas in the STEM fields, present contextual realities within Latina culture that apply to this specific group—as Latinas, as women, as hard science students, and for many, as immigrants or first-generation college students. Moreover, these aspects indicate that the culture of Latinas directly affects their resilience. Their cultural identity as Latinas can help them form resilience in the face of hardships and eventually find success by overcoming those hardships (Van Breda, 2001). The resulting sense of cultural pride provides a base of strength that further provides resilience (Morgan Consoli, Delucio, Noriega, & Llamas, 2015).

Context can come from the many types of exchanges that Latinas may experience. Interactions with others, dealing with stress, feeling a sense of community, feeling a sense of support, and understanding how they realize their potential are all factors that influence their resilience from this standpoint (Greene et al., 2004). Furthermore, involvement of a transactional dynamic process can be applied to Latinas in the STEM fields because the exchanges between the person and environment in the STEM fields are different for Latinas. Van Breda (2001) explained this as the *person-environment fit*, which takes into account the various systems an individual exists within and how they meet the demands of those various environments in order to succeed. Thus, it is also important to consider how contextual factors outside of individual characteristics function in how Latinas develop resilience (Morgan Consoli et al., 2015).

Finally, Greene et al. (2004) described resilience as "a multisystemic phenomenon that can occur across the life span" (p. 78). Latinas within the STEM field reveal experiences across the life spectrum with individuals, families, and communities that encourage unique paths of development. Latinas develop in this area by sharing their past and present experiences, including relation experiences with family, peers, and mentors (Rodriguez, Doran, Sissel, & Estes, 2019).

## Literature Review

### *Support of Latinas in Higher Education and STEM*

Various resilience factors assist Latinas in STEM majors, in both their studies and their entrance into STEM fields. Rendón, Nora, Bledsoe, and Kanagala (2019) proposed a Latinx student STEM success model that was comprised of six factors that provide support for Latinx students completing their STEM studies. These factors are (a) participation in STEM high-impact practices; (b) having multiple sources of financial support; (c) getting validation from significant others; (d) utilizing their own personal assets and ways of knowing; (e) becoming involved in Latinx-based STEM social and academic extracurriculars; and (f) using family cultural wealth learned at home. This model is a holistic approach to embracing Latinx student success within STEM by considering various avenues of support that all influence one another.

The first factor, participation in high-impact practices, has to do with becoming professionally involved with the STEM program of study. One form of professional development that helps Latinas is industry cooperation, like work-relevant trainings at community colleges (Santiago, Taylor, & Galdeano, 2015). The model shows the importance of STEM-focused social and academic areas of involvement. Becoming involved socially, academically, and professionally helps strengthen the sense of belonging within STEM studies, an issue that Latinas continually face (Martínez et al., 2019). As both women and Students of Color, Latinas are part of two marginalized groups that report the lowest sense of belonging in STEM (Rainey et al., 2018). Female students in engineering, for example, face a statistically significant higher risk of leaving engineering by their fifth semester than male students (Min, Zhang, Long, Anderson, & Ohland, 2011). In addition, for many Latinas/os, common cultural norms that drive attitudes and behavior, including degree completion, are further defined by gender (Cerezo, 2014), which is why broad forms of involvement starting before and during college is important for Latinas' persistence in STEM.

The strong sense of support that Latinas receive from involvement opportunities and environment are related to the acknowledgement of Latinas' culture. Latino culture places a high value on family consideration, and Latino families, when they are involved, positively influence the attainment of higher education, particularly in STEM (Hernandez, Rana, Alemdar, Rao, & Usselman, 2016). Latino cultural identity is highly important to Latinx students. The aspirations that Latino families have for their children to become family role models and succeed within another culture (Chlup et al., 2018) are part of the cultural wealth Latinas gain at home (Rendón et al., 2019). Cultural wealth fuels the personal assets and resilient characteristics Latinas carry with them in their efforts to succeed. Strengths founded in Latinas' cultural background have not garnered much research attention (Rendón et al., 2019), but they have been shown to provide aspirational capital that enforces resilience toward a better future (Yosso, 2005). The depths to which embracing Latinas' cultural identity can help ground them in order to succeed at the postsecondary level, let alone within a STEM department, is multifaceted. Whether in curriculum programming or in providing more mentors at the faculty level, which requires diversity initiatives at the institutional policy level, using

Latinas' cultural background as a tool to engage them with their educational environment is a necessary strategy.

### Latinas at an R1 Doctoral Hispanic-Serving Institution in Texas

Texas is one of the five US states with the largest Latinx student enrollment, but it is also one of the states with the largest completion gaps, −17%, between Latinx and White students (Excelencia In Education, 2019). At the university involved in this study, 38,597 undergraduate students were enrolled in the fall 2019 semester at the main campus—50% female (19,360) and 50% male (19,237). Students of Hispanic/Latino ethnicity are the largest minority group at the university, with 13,803 students making up 35.8% of the undergraduate student body.

In the fall of 2017, of the 1,065 ranked faculty at the university, 39 were Hispanic females and 45 were Hispanic males, together accounting for 7.9% of ranked faculty on campus. In the fall of 2019, of the 1,078 ranked faculty at the university, 42 were Hispanic females and 47 were Hispanic males, together accounting for 8.3% of ranked faculty on campus. Thus, the institution has shown an increase of Hispanic faculty members, providing more potential mentoring opportunities for Latina students by these new faculty members; however, the ratio between Latino students and faculty is still low.

### Methodology

An exploratory/descriptive qualitative case study (Yin, 2009) was used to generate a portrayal of the perceptions of Latina STEM students in an R1 Doctoral HSI in Texas. Hispanic-serving institutions must have a minimum enrollment requirement of at least 25% of the undergraduate students who identify as Latino/Hispanic, and the institution in this study acquired the designation of HSI in 2012. All data collection, analysis, and report writing for this study were based on qualitative research methods consisting of interviews and observations. These methods were conducted by the researchers in order to understand the complexity of the issue and bring forth overarching findings. To code and analyze the qualitative data, data were organized, sorted, and interpreted based on the participants' verbatim responses. Audio files and notes to compose a list of themes, categories, and sub-categories from the units of data (e.g., statements of meaning, quotes made in the participants' own words) were analyzed inductively. Institutional Review Board approval was obtained from the institution.

Qualitative research methods were used to analyze these data through the use of content analysis and constant comparative method (Lincoln & Guba, 1985; Gonzalez & Forister, 2015). The constant comparative method is an analysis that "involves systematically examining and refining the variations in emergent and grounded concepts" (Patton, 2015, p. 439) in order to develop categorized information in the form of themes or codes (Creswell,

2012). The categories that come about can assist in developing a deeper understanding of the data (Grove, 1988). Through this process, the researchers allow themes to emerge from the data instead of prescribing themes to the data. Once themes have emerged, the researchers can further engage the data in an inductive manner to connect codes and themes to the guiding theory (Creswell, 2012). In regard to this study, it was considered inductive because the current study hoped to link the experiences of Latinas in STEM as confirmation of and connection to the existing concept of resilience theory. In doing so, the constant comparative method allowed evidence to develop that tested the working hypotheses (Glaser, 1965).

This exchange process across the available data aligns with the constructivist nature of qualitative research, which seeks to understand the multiple truths that can exist within the data (Gonzalez & Forister, 2015). Connections across interviews arise that connect to the idea guiding the research. Different researchers may find different themes in the data as well, but the constant comparative approach argues for the necessity of pieces that slowly become a whole (Grove, 1988). Furthermore, the interview transcripts are treated as content that gives voice to "the lived experiences of individuals" (Gonzalez & Forister, 2015, p. 98).

Qualitative content analysis is "any qualitative data reduction and sense-making effort that takes a volume of qualitative material and attempts to identify core consistencies and meanings" (Patton, 2015, p. 790). This method begins early in the data collection process, which allows for reengagement of content as the concept develops in depth (Zhang & Wildemuth, 2009). This study applied content analysis to each of the interview transcripts to identify recurring words, phrases, and themes throughout. The meanings that developed emerged from patterns and themes in the text (Patton, 2015). This analysis method helps researchers gain deeper insight that goes beyond the text in the data. It allows for participants' stories to be understood from both their personal perspective and the researchers' scientific perspectives (Zhang & Wildemuth, 2009). Having both vantage points is important because the lived experiences of the participants are subjective stories that contain the scientific data used to help answer the guiding research questions.

Data for this research consisted of interviews and observations of ten undergraduate students in STEM fields. The researchers used an open-ended interview protocol that was expanded upon and revised as the research progressed. The use of an interview guide was combined with additional data being collected regarding demographic information for the interviewees (Patton, 2015).

## Findings

Themes and subthemes that emerged from the analysis of the data include *experiencing academic difficulty, acknowledging adversity, feelings of anxiety, feelings of isolation, statements of hope and/or positivity, experiencing lack of support, experiencing signs of support, losing hope/questioning, aspects of self-identity,* and *dealing with transitions.* Each of these themes relates to at least one of the three research questions guiding this study. The additional themes of *experiencing lack of support* and *losing hope/questioning* emerged during the content analysis, which highlights the necessity for further investigation into Latinas' resilience abilities to break through into STEM persistence and graduation success.

### Findings Related to Research Questions

*Research Question 1*

The first research question asked the following: How does resilience in Latina students—in relation to culture, background, and family support— play a role in their persisting toward the completion of a college STEM degree? Themes that emerged in relation to this research question are discussed next.

Within the context of the study, the theme of *aspects of self-identity* had a concept similar to experiencing signs of support due to homing in on the influences of the students' many identities, such as being STEM majors, Latinas, and females. The themes differ in that aspects of self-identity point to the various identities the students may strongly relate to in different situations (e.g., academic identity versus cultural identity), while experiencing signs of support are direct instances of students getting support from the different groups they identify with (e.g., support from those within their academic environment or from their cultural/familial environment). The *academic identity* subtheme revealed that many participants engaged with STEM topics from a young age, showing signs of curiosity that led to chosen challenges in their schoolwork. For these Latinas, being a STEM major was not arrived at through intimidation; instead, it often appeared to be fueled by curiosity. These STEM students were interested in STEM because they had a curious nature—some from a very young age, others from becoming more curious in college (perhaps by participating in STEM-related extracurricular activities). The *cultural identity* subtheme identified instances where Latinas related with cultural aspects of themselves as Latinas within their surrounding educational context, such as noticing Latinx academic professionals and faculty and other Latinas in their STEM classrooms or departments. This effect is an important element highlighted in the literature, which notes how peers and role models provide students with a source of hope for success within STEM for themselves in the future (Ruiz, 2013). The *gender identity* subtheme was similar, except it solely focused on when the Latina noticed particular female role models within STEM fields.

Within the theme of *acknowledging adversity*, we found instances where interviewees described adversity stemming from family (support/responsibility to family). Specifically, this adversity dealt with being responsible for their family unit. The acknowledgement of the support they felt they had to give back to their family in return added stress to their academic lives.

The theme of *experiencing signs of support* was categorized based on the environment; the subtheme *support from academic community* emerged from instances of Latinas getting direct support from advisors, professors, counselors, or any other academic professional. The subtheme of *support from peers/friends* included support that stemmed from classmates or close friends. The third subtheme was *family source of support/inspiration*. For some, inspiration came from family members interested in STEM, which is important in Latina STEM student development (Rodriguez et al., 2019). Significantly, the overarching theme of *experiencing signs of support* had a cultural dynamic and was also revealed in the literature to be an important factor in minority student success within STEM (Talley & Ortiz, 2017).

Identity formation embodies ethnic-specific values to characteristics learned from the participants' upbringing. Experiences from their upbringing helped Latina participants find a desire and reason to succeed when faced with challenges in the academic environment. The ability to transform their challenges was founded upon their cultural background and their personal experiences as Latinas. One participant, while discussing her hardships, reflected on her tough upbringing as both a Hispanic person in the United States and someone who was low-income and undocumented. She looked to her experiential cultural upbringing to find strength for persisting. Her personal experience provided a sense of strength to move forward and overcome previous identities that came with adversity, such as the financial difficulties of being low-income. Being labeled as different, both culturally as a Latina and legally as undocumented, provided the participant with inspiration to improve on her current and past situations.

For others, the *aspects of identity* theme blended with the theme of *acknowledging adversity*, which when combined can also form stories of resilience. Another participant pointed out three aspects of her self-identity: first-generation, Latina, and female. She expressed her disadvantaged place in society under these identities by acknowledging the opportunity that can arise from adversity. She found motivation within her culture and background to take on any opportunity that came her way. For many of these students, college is an opportunity that is new to both them and their families. This dynamic presents the possibility for family support to drive their persistence. Another participant shared, "I understand it's kind of hard for parents to understand what college is, but they just said they're proud and they're excited." Similar statements of family support from other participants revealed that even if the encouragement is solely verbal and reveals misconceptions about the college experience, the show of support is acknowledged. The participants understood that the family was with them in their current educational process, even if the process was unfamiliar to those family members. Thus, experiencing signs of support, specifically within the sub-theme of family source of support/inspiration, resonates strongly with this research question. The importance of the family unit in Latina persistence and success is evident across the literature as well as in this study's interviews.

Another influence that family support had on some of the participants came directly from other family members' direct involvement with STEM. This involvement inspired participants to pursue STEM disciplines as well, either because they witnessed a sibling doing well in the field or because they had family members already working in the field. This intrinsic aspect of family support was not available to all the Latinas interviewed, as they all had unique life experiences, but those who did have this influence expressed a strong urge to want to follow in the footsteps of those who inspired them. One Latina shared this sentiment in a story about her brother:

> When I was in high school, he was at college. And I will always see him [going to sleep] at two in the morning doing his homework, and also crying in his room trying to solve problems, but he couldn't because it was hard in the math, like, the math problems. So I always grew up thinking it's going to be hard for me. But I can do it. If my brother did it, then I can do it too.

Stories like the preceding example nourished resilience in Latina partici-pants and helped them develop a strong academic identity within STEM before they even entered STEM programs in college. The themes of *aspects of self-identity* (specifically the *academic identity* subtheme), *experiencing signs of support*, and *family source of support/inspiration* played a role in this aca-demic side of the participants' experiences. Some participants acknowledged possessing academic capabilities as both STEM students and academically capable students from an early age. They also attributed their interest and eventual capabilities in STEM to being inspired by family members who also took the STEM route. As referenced in the literature, the many iden-tities that Latinas develop through their academic experience help them better form their collegiate identity within STEM (Rodriguez et al., 2019). Acknowledging their academic capabilities fostered the belief that they could achieve at the higher-education level, even when considering what institutional type to attend. As one participant remarked, "I said no to com-munity college. I want to go to university."

*Research Question 2*

The second research question asked the following: How does social climate influence Latina students in the STEM fields? Related themes that emerged from this research question are discussed next.

Some participants did encounter transitions in their education path, such as moving from a community college to a four-year college. One participant, when reflecting on being a transfer student, identified the social aspect of relating with other transfer students, specifically another Latina, as helping to deal with change. She stated,

> It's funny cause I guess all the transfer students kinda . . . have classes that we're missing, so we kinda banded together. . . . Like, I feel like we have a group. We meet up sometimes too outside of school to just not think about engineering and just kinda, like, have fun. Yeah. One of them is also a Latina, so I feel like that helps because we're able to understand each other's struggles.

The same participant continued discussing this topic in the college context. Again, interestingly, the concept of diversity became an important topic to note. Another student shared how transitioning to a more diverse academic setting in middle school and high school was different for her, but she portrayed the transition in a positive light. Once the student started her STEM program in college, she noted the lack of diversity in her classes and also noted the lack of non-Hispanic diversity in her hometown:

> And it wasn't until I was in more diverse classes in my major that I felt like I really did belong. I guess that was a negative experience, but on the positive side, again, coming from a very predominantly Hispanic town . . . it was good for me to be exposed to so much diversity. I think that that really helped me.

Many multilayered themes from this study are reflected in this Latina's statement. Although she expressed concerns about coming from a nondiverse environment, she immediately turned to resilient thinking by making a statement of hope/positivity. In other words, she viewed her background experience of moving from a predominantly Hispanic setting to a more diverse setting as a beneficial event in her life.

Seeking a sense of belonging was an ordeal for various participants. For those participants who did not have the capital to take on the lack of diversity present in STEM, the experience could be more difficult to process. In an attempt to seek situational improvement, some Latinas try to immerse themselves in their STEM environment to feel more included. The resulting theme of *feelings of isolation* was manifested when one participant explained her negative experience when she tried to take part in STEM-related extracurriculars and did not feel welcome. Thus, the participant identified a negative aspect of the social climate that hindered her from feeling herself an equal peer within STEM. Because she already felt different due to being a Latina and undocumented, the transition to being isolated in her STEM studies because she was female increased her feelings of isolation.

These experiences that leave the participant feeling isolated lead to notions of feeling different. For Latinas in STEM, these situations, in comparison to other major fields, heighten the chances of departure without a degree or of switching one's major of choice (Riegle-Crumb, King, & Irizarry, 2019). The feelings of isolation, specifically from resulting negative peer interactions, can also trigger feelings of anxiety, especially when confronted with assignments such as team projects.

These experiences blended with the aspects of self-identity for some of the participants. Such experiences once again showed the layered benefits that can exist in the STEM environment when supportive structures, like equitable Latinx representation, are in place. Certain Latina participants sought STEM-related extracurricular involvement to boost their STEM academic identity. One participant highlighted the sense of belongingness in STEM while discussing an extracurricular research opportunity:

> I've been to other . . . organizations, . . . and the only one I've ever really felt comfortable with is theirs. And that's because with them I feel like it's like a collective space. We all have a shared experience as women in STEM and not like, . . . we're all from different races and stuff, but I don't know, I feel like we can relate to each other a lot more and have those conversations about like, oh, school and all that.

The participants made attempts to lean on their sources of support to push through the difficulties or to find new realms of support to better guide them. In the process, they elicited more streams of resilience that in turn helped them continue persisting toward their goal of degree completion. Along the path, these Latinas started to learn strategies that helped them deal with difficulties in the social climate that may present itself sometimes. These strategies are the focus of Research Question 3.

*Research Question 3*

The third research question asked the following: What types of successful strategies do Latinas develop in their pursuit of a college STEM degree? Related themes are discussed in the following paragraphs.

The theme *statements of hope and/or positivity* had three subthemes: *positive outlook/future plans, seeking/demonstrating resilience*, and *sense of belonging*. These three subthemes were crucial to this study because they included dialogue on Latinas' goals, facing and overtaking challenges, feelings of belonging in their STEM major, and the positive side to negative interactions. Having a positive attitude and making the decision to face challenges were both internalized methods incorporated into successfully pursuing a STEM degree. As one participant shared her background story regarding her college applications, she realized that she found a sense of self-reliance that pushed her forward. She stated,

> Ever since then, I just thought that I need to always be able to fix my own problems. I can't rely on not even my parents. . . . So, then, I would always be the one that got good grades in math and the good grades in science.

This anecdote on strength in upbringing reveals the participant's familial, navigational, and resistant capital (Yosso, 2005), all of which fuel resilience. She also showed a sense of courage to want to address her own issues. The stressful situations she faced with her family provided the opportunity for her to survive and recover from those situations. Another participant emphasized that her Latina identity fueled her positivity, stating, "Just being a Latina myself just makes me, like, push more and . . . wanting to keep going and pursuing, like, whatever comes next. It doesn't stop me."

The lessons learned by these students are then applied toward future stressful situations, like trying to achieve good grades in math and science courses. With resilience, Latinas have the capability to take charge and overcome demanding obstacles. If provided with the addition of a proper social climate comprised of guidance, support, networking, and self-efficacy, Latinas are better able to succeed in STEM fields because hope is instilled in the individual, and she is equipped to deal with hardships and face new obstacles. In describing her aspirations to succeed in STEM, one participant shared the following:

> I always knew I wanted to do something STEM related. I just didn't know what exactly, but I knew that I had to overcome all the difficult things that happened in my life to be that person. Be that STEM person that I wanted to be.

These positive attitudes become the supportive structure for a positive outlook that includes making hopeful academic plans, creating positive self-reflections, and experiencing growth. Another participant shared her reflection on this outlook:

> I think that moving forward with either graduate school or whatever comes next, I think that my experience has taught me to be more outspoken, more confident in what I have to bring to the table rather than feeling, like, oh, I got the imposter syndrome.

Skills and knowledge attained from facing adversities will, over time, provide participants with resilience (Van Breda, 2001) that can be used to overcome new academic challenges and transitions. Throughout her interview, one participant alluded to some instances when she experienced support and others when she did not. This anomaly was most evident in her discussion on interactions with peers and mentors. The participant showed appreciation for the network support of relatable peers. Not surprisingly, the research is rich with findings and suggestions for institutions to embrace peer mentoring to better assist Latinas in STEM (Rendón et al., 2019).

The road to positive strategies can be muddled with negative experiences as well. An important development from this research was the emergence of two themes that present internal challenges to Latinas in STEM: *experiencing lack of support* and *losing hope/questioning*. These aspects present a darker picture beneath the resilient attitude of these students, particularly because of the invasive nature of these themes throughout the other themes. Issues like feeling marginalized and experiencing difficult transitions present challenges to Latinas in STEM, and the toll these issue take on them over time affects their ability to persist. Internally sensing a despair, or questioning their abilities, or noting instances of missing support can all threaten the persistence of these students within a field of study needing their representation.

### Additional Findings: Latina Resilience—A Need to Fight Back

Multiple participants alluded to disillusionment, specifically when pursuing graduate studies. The themes of *losing hope/questioning* and *experiencing a lack of support* appeared across participant interviews. One participant described how she was previously interested in the thought of graduate school, but discouragement has caused her to lose interest. She stated, "After graduation, I wanted to go to grad school, but I feel like that's kinda been beaten out of me recently. I don't—I don't know yet." This issue deals with representation of Hispanic students in STEM graduate programs, as well as Latina representation in the STEM workforce. The theme of *experiencing lack of support* was predominantly strong when it came to faculty interactions and relationships. Multiple participants shared this concern, making statements such as the following:

*Participant A:*  I don't think I ever . . . I don't know if maybe it's my specific degree, I don't think I've ever actually had a one-on-one connection with any of the faculty members.

*Participant B:*  I think that [more faculty interaction] would benefit a lot of people and have more opportunities for students to connect with the faculty because I don't think that was offered a lot.

Participants also shared many statements where they felt themselves questioning their ability to persist, sharing concerns such as "That's another day that I thought about changing my major, because I'm like, 'If I'm not good at this, I'm not good at anything.'" The despair extended specifically into their capabilities as STEM students as well. One participant commented,

"Every day, there's days where I'm like, 'Man, I should have just decided to do business and something, or supply chain.'" This specific theme of *losing hope/questioning* permeated various spheres, including family relationships, as demonstrated by one participant's remarks:

> I thought, *I'm not cut out to be premed, I'm going to fail, I'm not going to do this. This is Chem 1 and I can't hack this.* Then I took my second exam and also failed it. I thought, *I can't do this, I'm going to change my major, my mom's going to be so sad.*

The themes that carry a negative tone provide a glimpse into the effects that problematic STEM environments create for Latinas. The notion of *losing hope/questioning* can instill a feeling that can last, impacting Latina STEM students beyond graduation and into their STEM career goals (Castellanos, 2018). Because a lack of representation at the faculty and professional level is an issue often discussed in the literature, feelings of *losing hope/questioning* and *experiencing lack of support* need to be confronted through Latina resilience.

## Discussion

This study strengthens understanding of the causes and effects of Latina resilience in the STEM fields by offering a firsthand student perspective. The findings reinforce the importance of understanding the interplay between culture, background experiences, and family support in helping Latinas persist and graduate within STEM. Previous studies on Latina identity formation by the principal investigator found that family, culture, and past life experiences significantly influence Latinas' retention and decisions to attend college, particularly as STEM majors (Gonzalez & Myers, 2016). Some preliminary results of this study confirm previous research findings but also indicate that participants from this study learned from obstacles by overcoming challenges and transforming them into successful experiences. These mechanisms all play a role in developing resilience in the Latina STEM college student and in how they demonstrate such resilience. The participants in the study provided examples that illustrate the positive aspects they find in being women, STEM students, and Latinas. These areas of inspiration are the ones that institutional leadership and policymakers need to tap into in order to better guide this population on a path of resilience and success.

Because of the burgeoning Latinx population in the United States, Latina college student success is essential for the nation's STEM future. Latinas continue to show heightened interest in STEM but are running into an established system and culture that does not historically support either females or minorities. Hispanic-serving institutions are trying to change that environmental pattern by trying to embrace their designation and becoming more purposefully *Hispanic-serving* (Garcia, 2019). As research trends have shown, how this designation translates into Latinx student success is a relevant question under investigation. This study provides a voiced direction to that question and a request for action to HSIs. Informing the research in regard to what Latinas are facing and feeling and how they are responding to the obstacles allows for a better understanding of the way Latinas interact in the existing social climates of higher education, especially in a male-dominated field. Through their stories, Latinas reinforce the ideas of persisting in STEM while facing difficulties presented to them due to gender, ethnicity, or unwelcomeness. Their backgrounds provide them with the desire to succeed and move up in life, while their academic experiences showcase their desire to mentor and be mentored, persevere, and be resilient.

## Recommendations for Policy and Practice

Castellanos (2018) suggested that institutional leaders should consider policies that support Latina involvement in academic programs and engagement through faculty mentorship relationships. Student–faculty relationships enhance accountability and have been shown repeatedly to be relevant in Latinx student success (Rodríguez & Oseguera, 2015). These policies should also include institutional support that takes both cultural and gender aspects into account, particularly in HSIs.

Institutional leadership should also look within their own ranks to improve the situation for females in STEM. Involving institutional leaders is essential in implementing change goals and advancing cultural change for women in STEM because they can voice concerns and can call the gender issues within STEM to attention (Austin, 2011; Rodríguez & Oseguera, 2015). Although the literature emphasizes persistence toward degree completion, the examination of Latina STEM majors' college experience is also vital for learning about new support measures administrators can take to successfully lead Latinas toward STEM careers (Castellanos, 2018). Empowering Latinas to feel capable in their math and science skills from an early age is also essential in improving their perceptions of self as STEM majors in college (Fouad, Santana, Lent, & Brown, 2017). With healthy social support via relatable role models, minority women like Latinas are better able to form their own self-identities as future workers in STEM (Fouad et al., 2017). Increasing the number of Latinas in STEM roles postgraduation is important for creating relatable role-model relationships—which is crucial because both women and Latinos, as separate groups, continue to be underrepresented in science and engineering education and employment (National Science Foundation, 2019).

In academia, Latinas as STEM faculty provide this same role-model dynamic. Research indicates that HSIs provide supportive STEM learning environments when they have a large concentration of Latinx faculty and peers, which results in strong faculty–student advising relationships (Revelo & Baber, 2018). Although seeing other Latinas find success in

STEM is important for the formation of these relationships, positive relationships with any faculty members can also help. Studies have shown that professors who encourage female students and create supportive relationships with them help those female students persist in their STEM major studies (Skolnik, 2015). With proper support, Latinas have shown equal or greater persistence rates than their Hispanic male peers (Fleming & McPhail, 2019). Tapping into this possibility is crucial for better serving Latinas in STEM. This consideration is vital when considering that Latinas, heretofore underserved as students and neglected as contributors to society, represent an integral piece of the surging Hispanic population and are thus a critical component of both the present and future of the United States.

## References

Ainsworth, M. D. (1989). Attachments beyond infancy. *American Psychologist, 44*(4), 709–716. https://doi.org/10.1037/0003-066x.44.4.709

Austin, A. E. (2011, March). *Promoting evidence-based change in undergraduate science education.* Retrieved from https://sites.nationalacademies.org/cs/groups/dbassesite/documents/webpage/dbasse_072578.pdf

Boutin-Martinez, A., Mireles-Rios, R., Nylund-Gibson, K., & Simon, O. (2019). Exploring resilience in Latina/o academic outcomes: A latent class approach. *Journal of Education for Students Placed at Risk (JESPAR), 24*(2), 1–18. https://doi.org/10.1080/10824669.2019.1594817

Castellanos, M. (2018). Examining Latinas' STEM career decision-making process: A psychosociocultural approach. *The Journal of Higher Education, 89*(4), 527–552. https://doi.org/10.1080/00221546.2018.1435133

Cerezo, A. C. (2014). Giving voice: Utilizing critical race theory to facilitate consciousness of racial identity for Latina/o college students. *Journal for Social Action in Counseling & Psychology, 5*(3), 1–24.

Chlup, D. T., Gonzalez, E. M., Gonzalez, J. E., Aldape, H. F., Guerra, M., Lagunas, B., . . . Zorn, D. R. (2018). Nuestros Hijos van a la Universidad [Our sons and daughters are going to college]: Latina parents' perceptions and experiences related to building college readiness, college knowledge, and college access for their children—A qualitative analysis. *Journal of Hispanic Higher Education, 17*(1), 20–40. https://doi.org/10.1177/1538192716652501

Creswell, J. W. (2012). *Educational research: Planning, conducting, and evaluating quantitative and qualitative research.* New York, NY: Pearson Education.

Excelencia in Education (2019). *Latinos in higher education: Compilation of fast facts.* Washington, DC: Excelencia in Education.

Fleming, J., & McPhail, I. P. (2019). *Success factors for minorities in engineering.* Oxford, United Kingdom: Routledge.

Fouad, N., Santana, M., Lent, R., & Brown, S. (2017). SCCT and underrepresented populations in STEM fields: Moving the needle. *Journal of Career Assessment, 25*(1), 24–39. https://doi.org/10.1177/1069072716658324

Garcia, G. A. (2019). *Becoming Hispanic-serving institutions: Opportunities for colleges and universities.* Baltimore, MD: Johns Hopkins University Press.

Glaser, B. G. (1965). The constant comparative method of qualitative analysis. *Social Problems, 12*(4), 436–445. https://doi.org/10.1525/sp.1965.12.4.03a00070

Gonzalez, E., & Forister, J. (2015). Conducting qualitative research. In J. Forister & D. Blessing (Eds.), *Introduction to research and medical literature* (4th ed., pp. 97–110). Charlotte, NC: Information Age.

Gonzalez, E., & Myers, J. (2016, April). *Latina college students in STEM fields: Stories of success in Texas* (Paper presentation). 2016 Annual Meeting of the American Educational Research Association (AERA), Washington, DC.

Greene, R. R., Galambos, C., & Lee, Y. (2004). Resilience theory: Theoretical and professional conceptualizations. *Journal of Human Behavior in the Social Environment, 8*(4), 75–91. https://doi.org/10.1300/j137v08n04_05

Grove, R. W. (1988). An analysis of the constant comparative method. *International Journal of Qualitative Studies in Education, 1*(3), 273–279.

Hernandez, D., Rana, S., Alemdar, M., Rao, A., & Usselman, M. (2016). Latino parents' educational values and STEM beliefs. *Journal for Multicultural Education, 10*(3), 354–367. https://doi.org/10.1108/jme-12-2015-0042

Lincoln, Y., & Guba, E. (1985). *Naturalistic inquiry.* Newbury Park, CA: Sage.

Martínez, A. J. G., Pitts, W., de Robles, S. L. R., Brkich, K. L. M., Bustos, B. F., & Claeys, L. (2019). Discerning contextual complexities in STEM career pathways: Insights from successful Latinas. *Cultural Studies of Science Education, 14,* 1079–1103. https://doi.org/10.1007/s11422-018-9900-2

McFarland, J., Hussar, B., de Brey, C., Snyder, T., Wang, X., Wilkinson-Flicker, S., . . . Hinz, S. (2017). *The condition of education 2017* (Report No. NCES 2017–144). National Center for Education Statistics.

Min, Y., Zhang, G., Long, R. A., Anderson, T. J., & Ohland, M. W. (2011). Nonparametric survival analysis of the loss rate of undergraduate engineering students. *Journal of Engineering Education, 100*(2), 349–373. https://doi.org/10.1002/j.2168-9830.2011.tb00017.x

Morgan Consoli, M. L., Delucio, K., Noriega, E., & Llamas, J. (2015). Predictors of resilience and thriving among Latina/o undergraduate students. *Hispanic Journal of Behavioral Sciences, 37*(3), 304–318. https://doi.org/10.1177/0739986315589141

National Science Foundation. (2019). *Women, minorities, and persons with disabilities in science and engineering: 2019* (Special Report No. NSF 19–304). National Center for Science and Engineering Statistics. Retrieved from https://ncses.nsf.gov/pubs/nsf19304/digest/about-this-report

Ong, M., Wright, C., Espinosa, L. L., & Orfield, G. (2011). Inside the double bind: A synthesis of empirical research on undergraduate and graduate women of color in science, technology, engineering and mathematics. *Harvard Education Review, 81*(2), 172–201. https://doi.org/10.17763/haer.81.2.t022245n7x4752v2

Patton, M. Q. (2015). *Qualitative research and evaluation methods* (4th ed.). Newbury Park, CA: Sage.

Rainey, K., Dancy, M., Mickelson, R., Stearns, E., & Moller, S. (2018). Race and gender differences in how sense of belonging influences decisions to major in STEM. *International Journal of STEM Education, 5*(1), 10. https://doi.org/10.1186/s40594-018-0115-6

Rendón, L. I., Nora, A., Bledsoe, R., & Kanagala, V. (2019). *Científicos Latinxs: The untold story of underserved student success in STEM fields of study.* San Antonio, TX: Center for Research and Policy in Education, The University of Texas at San Antonio.

114   *Elsa M. Gonzalez et al.*

Revelo, R., & Baber, L. (2018). Engineering resistors: Engineering Latina/o students and emerging resistant capital. *Journal of Hispanic Higher Education, 17*(3), 249–269. https://doi.org/10.1177/1538192717719132

Riegle-Crumb, C., King, B., & Irizarry, Y. (2019). Does STEM stand out? Examining racial/ethnic gaps in persistence across postsecondary fields. *Educational Researcher, 48*(3), 133–144. https://doi.org/10.3102/0013189x19831006

Rodríguez, L. F., & Oseguera, L. (2015). Our deliberate success: Recognizing what works for Latina/o students across the educational pipeline. *Journal of Hispanic Higher Education, 14*(2), 128–150. https://doi.org/10.1177/1538192715570637

Rodriguez, S. L., Doran, E. E., Sissel, M., & Estes, N. (2019). Becoming La Ingeniera: Examining the engineering identity development of undergraduate Latina students. *Journal of Latinos and Education.* https://doi.org/10.1080/15348431.2019.1648269

Ruiz, E. C. (2013). Motivating Latina doctoral students in STEM disciplines. *New Directions for Higher Education, 163*, 35–42. https://doi.org/10.1002/he.20063

Santiago, D., Taylor, M., & Galdeano, E. (2015). *Finding your workforce: Latinos in science, technology, engineering, and math (STEM): Linking college completion with US workforce needs: 2012–13.* Excelencia in Education. Retrieved from www.edexcelencia.org/research/workforce/stem

Skolnik, J. (2015). Why are girls and women underrepresented in STEM, and what can be done about it? *Science & Education, 24*(9), 1301–1306. https://doi.org/10.1007/s11191-015-9774-9776

Steele, H., & Steele, M. (1994). Intergenerational patterns in attachment. In K. Bartholomew & D. Perlman (Eds.), *Attachment in adulthood: Advances in personal relationships* (Vol. 5, pp. 93–120). Jessica Kingsley.

Stewart, M., Reid, G., & Mangham, C. (1997). Fostering children's resilience. *Journal of Pediatric Nursing, 12*, 21–31. https://doi.org/10.1016/s0882-5963(97)80018-8

Talley, K. G., & Ortiz, A. M. (2017). Women's interest development and motivations to persist as college students in STEM: A mixed methods analysis of views and voices from a Hispanic-serving institution. *International Journal of STEM Education, 4*(1), 1–24. https://doi.org/10.1186/s40594-017-0059-2

Van Breda, A. D. (2001). *Resilience theory: A literature review.* Pretoria, SA: South African Military Health Service.

Yin, R. K. (2009). *Case study research: Design and methods* (4th ed.). Newbury Park, CA: Sage.

Yosso, T. (2005). Whose culture has capital? A critical race theory discussion of community cultural wealth. *Race, Ethnicity and Education, 8*, 69–91. https://doi.org/10.1080/1361332052000341006

Zhang, Y., & Wildemuth, B. M. (2009). Qualitative analysis of content. In B. Wildemuth (Ed.), *Applications of social research methods to questions in information and library science* (pp. 308–319). Santa Barbara, CA: Libraries Unlimited.

# 7 First-Generation Latina Engineering Students' Aspirational Counterstories

*Tamara T. Coronella*

## Introduction

Mounting evidence supports the economic benefits of increasing and diversifying the science, technology, engineering, and math (STEM) workforce as a national strategy to maintain the United States' global competitiveness (U.S.; National Science Foundation, 2012). STEM jobs, most of which require a STEM undergraduate degree (Noonan, 2017), are among the fastest-growing careers (Carnevale, Porter, & Landis-Santos, 2015). A major concern is that the STEM workforce includes an overwhelmingly large proportion of White males, who form a shrinking demographic (Carnevale, Smith, & Melton, 2011). Thus, diversity in the STEM industry is not only desirable for the sake of inclusion but is also necessary to meet labor market demands. Minoritized[1] populations, therefore, need engagement that facilitates their inclusion in this prominent segment of business and industry (Carnevale et al., 2011; Ong, Wright, Espinosa, & Orfield, 2011).

One such population is the Latinx demographic, which, despite being the fastest-growing demographic in the United States, remains underrepresented in STEM careers (Chang, Sharkness, Hurtado, & Newman, 2014). While Latinx people comprise 17% of the total population, they hold only 11% of general engineering jobs (Carnevale et al., 2015). Women of all races, too, are underrepresented, as they hold only 24% of STEM jobs but constitute 47% of the overall working population (Beede et al., 2011). This imbalance is even greater when looking specifically at engineering jobs, only 13% of which are held by women (Society of Women Engineers, 2018).

Another population minoritized in STEM is the first-generation demographic. First-generation students, those individuals whose parents did not complete college, comprise approximately one-third of the overall college population (Cataldi, Bennett, & Chen, 2018). First-generation students enroll in college at lower rates than students whose parents graduated from college (Mitchall & Jaeger, 2018), are more likely to work while enrolled in college, and attend lower-performing schools (Cataldi et al., 2018).

Research has attributed minoritized student STEM departure to a multitude of factors. Individual reasons for departure, such as poor self-efficacy, self-concept, and attitude towards STEM, seem to be the primary focus (Else-Quest, Mineo, & Higgins, 2013). However, minoritized students often describe a hostile, unsupportive *climate* as their primary reason for departure (Gloria & Castellano, 2012; Samuelson & Litzler, 2016). Ong et al. (2011) defined the STEM climate as "the interpersonal relationships with other members of the local STEM communities and the cultural beliefs and practices within STEM that govern those relationships" (p. 192). To further explore the potential of an intervention to mitigate the climate that minoritized students describe as hostile and unsupportive, the present study focuses on first-generation Latina students (FGLS) in engineering disciplines. To address the need for diversity and to achieve a more balanced demographic representation, it is vital to implement systemic improvements to create a more supportive climate to better retain FGLS in STEM programs.

System improvements can be made through programs or services that promote belonging (Ong et al., 2011), such as academic advising, which serves as a central university support resource for students to guide them through the achievement of their academic and professional goals (O'Banion, 1994). Academic advising involves much more than simply providing advising services, especially if it is intended to promote success. Advisors should be responsive to students' needs, and by incorporating students' aspirations, assets, stories, and experiences into advising, they will be better positioned to support students. To capture some of these unique needs of FGLS in engineering, this study addressed the following research question: How do first-generation Latina students describe their goals and aspirations within the engineering climate?

This study took place within the College of Engineering (COE) at Southwestern University (SU), a large institution in the southwest that focuses on an access mission and measures its success by whom it includes and their outcomes (SU Charter, 2019). The COE is ranked in the top ten of undergraduate engineering degrees awarded nationally (Roy, 2018). Furthermore, the SU COE ranks thirteenth in terms of engineering degrees awarded to women and eighth in terms of engineering degrees awarded to Hispanics nationally (Roy, 2018). These rankings demonstrate that there is room for improvement. In response to these persistent negative outcomes nationally and within SU, this study considers how an academic advising intervention can support the aspirations of FGLS to complete a degree in engineering.

## Relevant Literature

The following literature review begins with a discussion of academic advising, followed by a discussion of the STEM climate. These discussions consider issues related to the race, generational status, and gender of students. This literature review is supported by the theoretical framework and models of validation (Rendón, 1994), community cultural wealth (CCW; Yosso,

2005), and *deficit viewpoints* (Valencia, 2010). Finally, validating advising practices are described.

## Background

### Academic Advising

Academic advising is a university-provided resource designed to support student persistence (Pascarella & Terenzini, 2005). Academic advisors, whether professional staff or faculty, work with students holistically to collaborate on the students' completion of their academic and professional goals (O'Banion, 1994). Advising practices are rooted in student development theories (McGill, 2016), which are programmatic in nature, "based on what professionals do to encourage learning and student growth" (Patton, Renn, Guido, & Quaye, 2016, p. 8). Research has shown a positive correlation between successful advising and persistence (Pascarella & Terenzini, 2005), but there is a lack of literature on variations within advising practices that can better support minoritized students across STEM disciplines and specifically in engineering (Ong et al., 2011).

Many student development theorists developed generalizable frameworks (Hernández, 2017) that minimally focused on the role of gender, race, or ethnicity in advising practices (Pascarella & Terenzini, 2005; Patton et al., 2016); therefore, developmental advising is rooted in an exclusionary theoretical framework. To provide a truly holistic approach, student support resources must incorporate the experiences and identities of minoritized students, which leads to an engagement of social justice and political issues (Patton et al., 2016).

### The Racialized and Gendered Climate

In the context of higher education, a negative racial climate is a "social and academic environment that exhibits and cultivates racial and gender discrimination" (Yosso, 2006, p. 101). Specifically, Latinas in STEM experience racial battle fatigue (Smith, Yosso, & Solórzano, 2011), race and gender microaggressions (Yosso, Smith, Ceja, & Solórzano, 2009), stereotype threat, and low faculty expectations (Solórzano, Villalpando, & Oseguera, 2005) as hurdles impeding degree attainment. These negative experiences reinforce exclusionary outcomes. Indeed, a well-documented explanatory factor for minoritized students' low participation rates is the racialized and gendered climate in STEM (Samuelson & Litzler, 2016).

Prior research has cited successful practices to address the racialized and gendered climate, such as those that foster inclusion (Yosso et al., 2009), develop women's sense of identity (Blackburn, 2017), and align with societal and familial oriented goals (Diekman, Weisgram, & Belanger, 2015). Addressing minoritized students' negative experiences by developing strategies to ease their experience and help them achieve degree attainment should be a central focus in advising FGLS through their engineering studies. Rincón and George-Jackson (2016) found that within engineering, increased support systems like academic advising positively influenced underrepresented students' perceptions of the racialized and gendered climate.

## Theoretical Framework and Key Constructs

### Theory of Validation

The theory of validation emerged in the early 1990s from a study of the Transition to College Project, which considered how student learning was affected by student involvement in academic and social experiences (Rendón, 1994). Rendón (1994) reported five key findings that emerged from the research: (1) students expressed concerns about their ability to succeed; (2) intervention from key college representatives can help to ameliorate these concerns; (3) success in college depends on an "external agent [who] can validate [minority] students in an academic and interpersonal way" (p. 8); (4) validation transforms students into powerful learners; and (5) "validation may be the missing link to involvement, and may be a prerequisite for involvement to occur" (p. 9). Validation occurs through interactions with an agent; academic advisors are considered *out-of-class validating agents*, and they can play a key role in providing academic and interpersonal validation (Rendón, 1994; Rendón Linares & Muñoz, 2011, p. 21).

When engaging a validating advising approach, the academic advisor seeks to empower, support, and affirm the student. Validating advising practices require advisors to actively participate in affirming their students' experiences and histories as forms of knowledge, assets, and strengths (Rendón Linares & Muñoz, 2011) to empower them to be successful. By incorporating the experiences and identities of individuals, advising is inclusive of students' and advisors' lived experiences (Hernández, 2017). Through a practice based on these validating approaches, academic advisors build a holistic understanding of the student's past and current experiences, which helps them better support their advisees. Students who experienced validation from an out-of-class institutional agent, such as an academic advisor, ultimately persisted at higher rates (Hurtado, Ruiz Alvarado, & Guillermo-Wann, 2012).

### Community Cultural Wealth

Central to validation is the incorporation of a student's strengths and assets. An asset-based approach, such as Yosso's (2005) community cultural wealth (CCW) model, considers how certain types of capital are developed within and from students' communities, becoming sources of strength and ultimately resources for mobility and advancement (Samuelson & Litzler, 2016;

Yosso, 2006). These overlapping and intertwined forms of capital include aspirational, navigational, social, linguistic, familial, and resistant capital (Yosso, 2005). Aspirational capital is one's ability to be hopeful about the future in the face of real challenges. Linguistic capital consists of intellectual and social abilities to communicate through and in multiple languages. One's sense of community and family nurtured among communities and families is familial capital. Social capital includes the community's resources, spaces, and networks, while navigational capital is the ability to operate within social institutions and places. Finally, Yosso and Burciaga (2016) described resistant capital as skills and knowledge that emerge by challenging systemic invalidations. Studies have explored the positive benefits of and increased need for validation of the CCW model within engineering classrooms and support services (Romasanta, 2016; Samuelson & Litzler, 2016), to increase the inclusion of and positive academic experience for minoritized students.

A long-standing, widely referenced model used to explain minoritized students' academic performance is *deficit thinking* (Valencia, 2010). This model attributes students' failures to their own internal deficits resulting from their family socialization, genetics, class, and culture (Valencia, 2010). Cultural capital, the most widely referenced capital in literature describing minoritized student departure (Solórzano & Yosso, 2002), comprises the system of meanings and symbolism within a community or group acquired by adopting the dominant group's values and norms (Lin, 2002). Education support services and programming focus primarily on helping students acquire the dominant cultural capital, thereby devaluing and invalidating a minoritized student's multitude of other capitals (Ong et al., 2011). By describing students as deficient, researchers argue that educators begin to expect students to be deficient (Martin, Smith, & Williams, 2018), thereby reinforcing low expectations and ultimately lower outcomes (Liou, Martinez, & Rotheram-Fuller, 2016). Bearing this in mind, educators are encouraged to consider their own expectations of and for students to better support positive outcomes (Martin et al., 2018).

Asset-based views serve as a counterpoint to deficit models of student support. Based on their study of how minoritized students experience STEM college programs, Beals and Ibarra (2018) concluded that the integration of student identities and their diverse ways of thinking better supports persistence. Thus, educational reform should focus on the forms of capital these students do have, rather than the capital they lack (Solórzano & Yosso, 2002). An asset-based approach accomplishes this by considering how varied types of capital that develop within and from minoritized communities can become sources of strength to support student persistence (Samuelson & Litzler, 2016). Stevenson et al. (2019) described how Latinas used their strengths to reject deficit viewpoints and were able to develop resiliency within the STEM climate as a result, such resiliency representing an individual's ability to succeed despite difficulties (Benard, 2004). Thus, validating advising practices in the context of this study engage asset-based views of students and focus on the activation of student strengths within the hostile climate of engineering.

## Method and Data Analysis

To achieve this study's aim of exploring how FGLS describe their goals and aspirations within the engineering climate, the participating FGLS shared their *testimonios*, or counterstories, in an action research, qualitative study. A testimonio is "a verbal journey of a witness who speaks to reveal the racial, classed, [and] gendered . . . injustices they have suffered as a means of healing, empowerment and advocacy" (Pérez Huber, 2009, p. 644). Testimonios are revealed through counterstories, which are used to capture the experiences of minoritized groups as legitimate data (Solórzano & Yosso, 2002). The data here were extrapolated from a larger action research study that captured validating advising interactions between academic advisors and their advisees (Buss, 2018).

The study was conducted at SU within the COE during the FGLS' first semester. Three purposively selected advisors participated. Ten FGLS were identified for participation through a stratified sampling procedure. All first-year students at SU are assigned an academic advisor; FGLS in the first semester of their engineering studies, assigned to the three participating advisors, were invited to participate. Two advisors worked with three student participants each, while one advisor worked with four students. Each advisor engaged in two individual meetings with each of her assigned students following a scripted set of narrative interview prompts designed to elicit the FGLS' forms of CCW. Narrative interviews are useful in eliciting meaning and knowledge through stories shared between the researcher and subject (Brinkmann & Kvale, 2015). Interview data were collected and analyzed. Additionally, advisors completed a template form (created by the researcher), noting each time FGLS discussed a form of CCW. Finally, the students participated in a focus group designed to gather their insights as part of the analysis process, which is further described in the subsequent section. The focus group was recorded and transcribed, and the resulting data were analyzed.

The data in this study were analyzed with two approaches—a coding process (Saldaña, 2016) and a three-phase *collaborative analysis process* (Pérez Huber, 2009), one process embedded within the other. The data corpus was first coded inductively, using initial coding, which is often considered a starting point for identifying patterns (Saldaña, 2016). The second component of the analysis incorporated use of the collaborative analysis model, which originated from Pérez Huber (2009), who utilized *testimonios* as a methodology to capture student stories. This approach incorporated student voices into both the data collected and the analysis of the data. During the focus group, students were provided with purposively selected statements from their fellow students, gathered during the students' individual meetings

*Table 7.1* Students' Educational Majors.

| Student Name (Pseudonyms) | Engineering Major |
| --- | --- |
| Emma | Environmental Engineering |
| Jenna | Environmental Engineering |
| Jessica | Environmental Engineering |
| Camila | Civil Engineering |
| Angie | Computer Science Engineering |
| Carson | Computer Science Engineering |
| Mona | Computer Engineering |
| Cindy | Biomedical Engineering |
| Emily | Biomedical Engineering |
| Marissa | Biomedical Engineering |

with their advisors. These statements were identified from the inductive round of coding and referenced the FGLS' gendered, racial, and generational status experiences. The focus group participants reflected upon those statements; their reflections helped to shape the analysis and understanding of the statements. In other words, participants were engaged in the analysis of their own data. Their interpretations were captured in the broad data corpus and included in the overall analysis. From there, a round of thematic coding was conducted to identify the FGLS' forms of capital (Saldaña, 2016). Therefore, the students shared and shaped their stories to inform as a form of counterstorytelling.

## Key Findings

In the following section, I discuss the study's findings in the sequence of (1) familial aspirations, (2) communal goals, and (3) activation of aspirations in response to the climate. The findings are followed by recommendations on ways that universities can develop policies, norms, and practices to further support the educational aspirations of FGLS in engineering based on those findings.

### Engaging Familial Aspirations

As discussed, deficit viewpoints undermine reform efforts for minoritized students and disregard their wealth of capital (Rendón, 1994, 2002; Yosso, 2005). These FGLS students had arrived at college with optimism about their success and clearly articulated aspirations. They expressed their long-standing, deeply ingrained intentions to attend and succeed in college. Mona described how her family "drilled" into her the expectation that she would complete a degree. She described her sister reading to her when she was a child and noted that her sister "is the reason I am going to college." Camila described how her father, a construction worker, would take her to work with him, which led to her interest in civil engineering. Angie's father directed her to "focus on school and homework" while not permitting her to hold an outside job during high school. These statements reflect their aspirational capital and hope that they would eventually attend college (Yosso, 2005). The students' statements are consistent with the findings of Chang et al. (2014) on how familial support can act as a mediator within the racial climate of STEM.

Multiple participants described the family support they received that promoted their attendance and motivated them to complete college. Emily revealed that her mother worked at a fast-food restaurant and told her "you have to better yourself, go to college." Marissa felt inspired by her grandmother, who had had to drop out of school and directed Marissa to complete her education. In addition to meeting familial pressures to accomplish a goal that her grandmother never had the opportunity to achieve, Marissa felt compelled to provide for her father. Thus, Marissa wanted to succeed to help provide for her family. She explained, "My dad didn't have a college degree . . . so right now he's working at a job where . . . they don't have a retirement plan for him or anything . . . he's just working until he can't anymore." The families of these FGLS helped them, directly or indirectly, to recognize the economic advantage of pursuing a degree.

These statements reflect the family's role in fostering and supporting the student's aspirational capital through their own aspirations for the student's success. Yosso (2005) described how the shared stories within a family "nurture a culture of possibility" for students to move beyond their parents' economic and vocational outcomes (p. 78). Indeed, "parental encouragement for their children's educational aspirations is one of the most important factors impacting those decisions to pursue higher education" (Mitchall & Jaeger, 2018, p. 583). Thus, while their parents did not attend college, the students internalized their parents' aspirations for their children to earn a degree, which fostered the students' aspirations (Mitchall & Jaeger, 2018). Cardoso and Thompson (2010) described how parents and their students "often make decisions that promote the social and economic stability of the entire family" (p. 260). In their research, they noted that family relationships emerge as a form of *protective factors*, which foster resiliency among minoritized students (Cardoso & Thompson, 2010).

### Engaging Communal Goals

The students described, too, how they pursued a degree in engineering as a function of their goals for society. Emma described how she was "really passionate about the environment." She was enrolled in a course that engaged with local community partners to identify solutions for a problem. Cindy expressed her desire to design prosthetics "cheaper and source out to different countries who can't afford [them]."

Beyond improving their community, these participants demonstrated a desire to inspire others coming behind them, which engages their social capital to continue the tradition of "lifting as we climb" (Yosso, 2005, p. 80). Cindy said, "I see becoming an engineer, a professional, someday becoming somebody respectable that is inspiration [sic] to them to keep fighting for something that they are not being taught to fight for." Cindy's comments reflect the reputational value of being an engineer and emphasize her belief that she would be more capable of enacting change with an engineering degree. This highlights her resistant capital as fuel for her engineering aspirations.

Overall, these statements reflect the FGLS' *communal goals* to care for and work with others (Diekman et al., 2015). Indeed, research supports the engagement of communal goals within the classroom and support services as a persistence strategy for women specifically (Blackburn, 2017). Students who have been involved with their communities and families report a greater sense of belonging in the college community, as they align their aspirations to their goals (Diekman et al., 2015). This alignment further supports what Benard (2004) described as a *sense of purpose*, which is a characteristic of resiliency. Research describes how Latinas mitigate the STEM climate through the activation of resiliency, as resilience predicts successful academic achievement (Morgan Consoli, Delucio, Noreinga, & Llamas, 2015).

### Activating Aspirations in the Engineering Climate

In addition to clearly articulating their goals, the FGLS described an increased, heightened consciousness of their gender, race, and generational status. Mona described her experience as a Latina in engineering in the following way: "I don't think of race as a factor in anything. I'm just a person, and it is what it is. . . . I don't ever really think of myself as an outsider because someone else was a different race than me." In this statement, Mona shared how she was not cognizant of her race playing a role in her experience. However, she subsequently commented, "But I did notice that some people tend to stick within the same race groups." This seeming contradiction carried through in her continued discussion. For example, she described a club she had recently joined that visits nearby towns and holds panels for first-generation high school students. Mona described her excitement, saying "When they told me that we were speaking to a lot of first-generation families . . . that's when it hit me, and I was like, Oh! I'm a minority person going to . . . I'm pursuing a degree that's dominated by men."

Through this comment, Mona displayed her intersectional identities as a first-generation Latina engineering student, operating as a woman in a male-dominated field with expertise she could share with other first-generation Latinas. Mona's next comments illustrate that she had realized her power:

I was like, "Wow! Oh." And that also ties into the whole sense of community because they wanted minorities to speak to minorities. Not like, a person whose families have had generations of college degrees going to speak to someone who's never even. . . . They never had the opportunity to go to college.

Mona is describing various forms of capital here, citing her navigational capital that enabled her to serve as a role model, her resistant capital that she could use to help lift up Latinas in the pipeline behind her, and her social capital, reflected by joining clubs and engaging with her peers. Angie described a similar realization about her resistant capital: "You take a step back sometimes and you look at it and it's like, 'Oh wait, yeah. A Latino woman did this and that's pretty cool' and 'That's me' and 'Maybe I'm a role model for someone else that's also trying'." Angie described the sense of pride she experienced as a Latina role model.

Cindy provided a stark description of her racialized experiences at SU. She said her race

comes into play when I want to ask for help from other people. Because, obviously, they'll have their groups, so they don't really know what to do when I want to join them. It shouldn't be that way, but I've seen that.

When probed to describe her experience further, she explained how she encountered a microaggression during an icebreaker activity in her residence hall:

So then, when we were doing the icebreaker, most of them went one way, and they just left me and these two other ethnic people. So, that was really awkward, but I tried to not let that get to me, because I have my goals that I want to get to. And, even if nobody wants to be with me, I'm going to get there some way.

Cindy's descriptions reinforce research that describes the microaggressions that Latinx students experience within the STEM climate (Rendón, Nora, Bledsoe, & Kanagala, 2019). Cindy stated, "I just feel like it's a big opportunity, because being part of a minority, so then if I keep pushing myself, and I'm doing great. You know people usually don't expect much of minorities." Cindy specifically cited a deficit-based view: People do not hold high expectations for minorities. However, Cindy described her resistant capital as she overcame microaggressions and remained steadfast in her engineering aspirations.

The participants provided the following descriptions of their experiences as women in the engineering program. Angie wished the engineering program included more women: "You just need more . . . women in the program. And that would just be welcoming by looking around you and seeing more people that." She additionally described how in her introductory engineering course of 40 students, she was one of only four women. She professed, "I don't see any of them, because they're like in the corners . . . it's just a bunch of like guy heads in front of me." In her comments she expounded upon the loneliness and isolation of being a woman in engineering.

Jessica's viewpoints were consistent with Angie's. She described her own introductory engineering course: "It was kind of weird at first being in a group with just all guys. . . . It was kind of intimidating because I didn't want to seem dumb or whatever." This experience directly reflects a gendered fear of confirmation of a stereotype about women. This evidence points to the literature that describes the isolation, fear, and loneliness women can experience in engineering (Samuelson & Litzler, 2016). Jenna declared, "Well, I feel a little tired. There are four in my group and I'm the only girl . . . sometimes they just all ignore me."

Alternatively, Emma found motivation from negative interactions with her male peers. She said, "I think like what makes me feel better is like if I'm doing well compared to the class . . . it's like 'oh, I'm just as good as [the men in class]'." Camila concurred with Emma: "Like she said, we're equally. . . . Basically, we can be as smart as them. We can apply ourselves as much as they do." Both Emma and Camila drew upon their resistant capital to combat stereotypes and remain focused on their academic goals (Yosso, 2005).

The participants described their experiences as first-generation students consistently. Carson said, "I think that the biggest difference is if you're not first-generation, you have people in your life that have gone to college and who completed college so it's not a big deal if you go." Angie shared how she has been "figuring stuff out by" herself. Mona's, Carson's, and Angie's comments reflect how they reinforced their navigational capital in college. Navigational capital emerges when individuals maneuver systems not created for them (Yosso, 2005), which these students have done throughout their educational experience. As described previously, deficit-based models emphasize what first-generation students lack, but these participants reflected a counterpoint as they expressed confidence about maneuvering higher education. This finding is consistent with research describing how the retention of high aspirations within difficult contexts fosters resilience to succeed (Perez Huber & Cueva, 2012; Romasanta, 2016).

## Discussion

The findings reveal how the participating FGLS arrived at SU with a depth of capital, interwoven assets, and goals that fueled their motivation to pursue engineering. Furthermore, they drew upon their identities and found strength within them as they navigated microaggressions and stereotypes experienced within the engineering climate. The students articulated communal goals for their families and communities, while drawing upon their resistant, familial, social, and aspirational forms of capital.

Overall, the students' statements reinforce findings regarding a form of CCW to complement Yosso's (2005) model: pluriversal (Rendón et al., 2019). A pluriversal form of capital is a Latina's ability to hold multiple meanings and consciousnesses simultaneously and draw upon that as a strength. Their strength is derived from multiple modes or even conflicting identities (Rendón et al., 2019). As women in a male-dominated field, they found fuel to persist. As Latinas in a predominantly White field, they held on to their aspirations.

Additionally, their increased consciousness and aspirations supported their internalized beliefs and drive to succeed. This could also be described as a sense of *autonomy* (Benard, 2004), defined as "an ability to act independently and feel a sense of control over one's environment" (Benard, 2004, p. 20). They described the agency they felt in being able to shape their careers and life outcomes through their aspirational and navigational capital. Autonomy is one of four main categories of resiliency, which refers to the activation of innate capacities to succeed despite barriers, such as a negative racial climate (Benard, 2004). This aligns with Romasanta's (2016) findings that FGLS develop their academic resiliency by engaging forms of CCW. Essentially, FGLS operationalize their forms of CCW as resiliency capital (Romasanta, 2016). Thus, this study further supports the connection between CCW and resiliency through the cultivation of assets within a hostile climate.

Finally, the assets became apparent to advisors through the advising interventions, which were designed to capture counterstories. Developmental advising definitions often describe what an advisor should discuss in a meeting as an advisee's goals, plans, and interests (O'Banion, 1994). However, this fails to capture the nuance of the types of discussions that *should be* occurring for minoritized students. Spaces that promote empowerment and resilience emerge as critical factors for academic and lifelong success (Benard, 2004; Perez Huber & Cueva, 2012). Therefore, rather than focusing on remediation through programming, colleges should instead explore how asset-based frameworks can be employed to reframe existing programming.

## Conclusion

The sustained underrepresentation of Latinas in engineering threatens economic vitality, innovation, and technology advancement in the United States (Carnevale et al., 2011). Furthermore, it sustains a deficit-based system that continues to maintain inequities based upon meritocratic perspectives. Alternatively, validating advising practices promote an inter-action that elicits the unique strengths and assets that Latinas possess. These assets serve as capital that can be activated to succeed within the hostile climate of engineering. Through their resilience, the students described a form of resistant capital. This study, therefore, supports the interplay of CCW and resiliency while adding another form of capital to the dynamic set of assets previously identified. Furthermore, students' goals and aspirations can be elicited through validating advising practices to create a more supportive climate in engineering, with the goal of increasing the number of STEM graduates.

## Note

1. I use the term "minoritized" instead of "minority" to acknowledge that individuals are not born into minority status, but "instead, they are rendered minorities in particular situations and institutional environments that sustain an overrepresentation of whiteness" (Patton, Harper, & Harris, 2015, p. 212).

## References

Beals, R. A., & Ibarra, R. A. (2018). Multicontextuality: A framework for access and assessment of underrepresented STEM students. In A. de los Santos, L. I. Rendón, G. F. Keller, A. Acerado, E. M. Bensimón, & R. J. Tannenbaum (Eds.), *New directions: Assessment and preparation of hispanic college students*. Tempe, AZ: Bilingual Press.

Beede, D., Julian, T., Langdon, D., Mckittrick, G., Khan, B., & Doms, M. (2011). *Women in STEM: A gender gap to innovation* (Issue Brief No. 04–11). Washington, DC: Economics and Statistics Administration.

Benard, B. (2004). *Resiliency: What we have learned*. San Francisco, CA: WestEd.

Blackburn, H. (2017). The status of women in STEM in higher education: A review of the literature 2007–2017. *Science and Technology Libraries*, *36*(3), 235–273.

Brinkmann, S., & Kvale, S. (2015). *Interviews: Learning the craft of qualitative research interviewing*. Los Angeles, CA: Sage.

Buss, R. R. (2018). Using action research as a signature pedagogy to develop EdD students' inquiry as practice abilities. *Impacting Education: Journal on Transforming Professional Practice*, *3*, 23–31.

Cardoso, J. B., & Thompson, S. (2010). Understanding the concept of resilience in Latino immigrant families. *Families in Society: The Contemporary Journal of Social Services*, *91*, 257–265.

Carnevale, A. P., Porter, A., & Landis-Santos, J. (2015). *Hispanics: College majors and earnings*. Center on Education and the Workforce. Retrieved from https://cew.georgetown.edu/projects/diversity/

Carnevale, A. P., Smith, N., & Melton, M. (2011). *STEM: Science, technology, engineering, mathematics*. Center on Education and the Workforce. Retrieved from https://cew.georgetown.edu/cew-reports/stem/

Cataldi, E. F., Bennett, C. T., & Chen, X. (2018). *First-generation students: College access, persistence, and postbachelor's outcomes* (NCES2018–421). Washington, DC: National Center for Education Statistics.

Chang, M. J., Sharkness, J., Hurtado, S., & Newman, C. B. (2014). What matters in college for retaining aspiring scientists and engineers from underrepresented racial groups. *Journal of Research in Science Teaching*, *51*(5), 555–580.

Diekman, A. B., Weisgram, E. S., & Belanger, A. I. (2015). New routes to recruiting and retaining women in STEM: Policy implications of a communal goal congruity perspective. *Social Issues and Policy Review*, *9*(1), 52–88.

Else-Quest, N. M., Mineo, C. C., & Higgins, A. (2013). Math and science attitudes and achievement at the intersection of gender and ethnicity. *Psychology of Women Quarterly*, *37*(3), 293–309.

Gloria, A. M., & Castellanos, J. (2012). Desafíos y bendiciones: A multiperspective examination of the educational experiences and coping responses of first-generation college Latina students. *Journal of Hispanic Higher Education*, *11*(1), 82–99.

Hernández, E. (2017). Critical theoretical perspectives. In J. H. Schuh, S. R. Jones, & V. Torres (Eds.), *Student services: A handbook for the profession* (pp. 205–219). San Francisco, CA: Jossey-Bass.

Hurtado, S., Ruiz Alvarado, A., & Guillermo-Wann, C. (2012). Inclusive learning environments: Modeling a relationship between validation, campus climate for diversity, and sense of belonging. In *Annual conference of the association for studies in higher education* (Vol. 53, pp. 1689–1699). Las Vegas, NV: ASHE.

Lin, N. (2002). *Social capital: A theory of social structure and action*. Cambridge, MA: Cambridge University Press.

Liou, D. D., Martinez, A. N., & Rotheram-Fuller, E. (2016). "Don't give up on me": Critical mentoring pedagogy for the classroom building students' community cultural wealth. *International Journal of Qualitative Studies in Education*, *29*(1), 104–129.

Martin, G., Smith, M., & Williams, B. (2018). Reframing deficit thinking on social class. *New Directions for Student Services*, (162), 87–93.

McGill, C. M. (2016). "Cultivating ways of thinking": The developmental teaching perspective in academic advising. *New Horizons in Adult Education & Human Resource Development*, *28*(1), 50–54. http://doi.org/10.1002/nha3.20131

Mitchall, A. M., & Jaeger, A. J. (2018). Parental influences on low-income, first-generation students' motivation on the path to college. *Journal of Higher Education*, *89*(4), 582–609.

Morgan Consoli, M. L., Delucio, K., Noreinga, E., & Llamas, J. (2015). Predictors of resilience and thriving among Latina/o undergraduate students. *Hispanic Journal of Behavioral Sciences*, *37*(3), 304–318.

National Science Foundation, Division of Science Resources Statistics (2012). Women, minorities, and persons with disabilities in science and engineering (Special Report NSF 11–309). Arlington, VA. Retrieved from www.nsf.gov/statistics/wmpd/sex

Noonan, R. (2017). Women in STEM: 2017 Update (ESA Issues Brief #016–17). Retrieved from https://eric.ed.gov/?id=ED590906

O'Banion, T. (1994). An academic advising model. *NACADA Journal*, *14*(2), 10–16. Reprinted from the *NACADA Journal*, *14*(2), 10–15.

Ong, M., Wright, C., Espinosa, L. L., & Orfield, G. (2011). Inside the double bind: A synthesis of empirical research on undergraduate and graduate women of color in science, technology, engineering, and mathematics. *Harvard Educational Review*, *81*(2), 172–208.

Pascarella, E. T., & Terenzini, P. T. (2005). *How college affects students: A third decade of research*. San Francisco, CA: John Wiley & Sons, Inc.

Patton, L. D., Harper, S. R., & Harris, J. (2015). Using critical race theory to (re) interpret widely studied topics related to students in U.S. higher education. In A. M. Martínez-Alemán, B. Pusser, & E. M. Bensimon (Eds.), *Critical approaches to the study of higher education: A practical introduction* (pp. 193–219). Baltimore, MD: Johns Hopkins University Press.

Patton, L. D., Renn, K. A., Guido, F. M., Quaye, S. J. (2016). *Student development in college: Theory, research, and practice.* San Francisco, CA: John Wiley & Sons, Inc.

Pérez Huber, L. (2009). Disrupting apartheid of knowledge testimonio as methodology in Latina/o critical race research in education. *International Journal of Qualitative Studies in Education, 22*(6), 639–654.

Perez Huber, L., & Cueva, B. M. (2012). Chicana/Latina testimonios on effects and responses to microaggressions. *Equity and Excellence in Education, 45*(3), 392–410.

Rendón, L. I. (1994). Validating culturally diverse students: Toward a new model of learning and student development. *Innovative Higher Education, 19*(1), 33–51.

Rendón, L. I. (2002). Community college *puente*: A validating model of Education. *Educational Policy, 16*(4), 642–667.

Rendón, L. I., & Muñoz, S. M. (2011). Revisiting validation theory: Theoretical foundations, applications, and extensions. *Enrollment Management Journal, 5*(2), 12–33.

Rendón, L. I., Nora, A., Bledsoe, R., & Kanagala, V. (2019). *Científicos Latinxs: The untold story of underserved student success in STEM fields of study.* San Antonio, TX: Center for Research and Policy in Education, The University of Texas at San Antonio.

Rincón, B. E., & George-Jackson, C. E. (2016). Examining department climate for women in engineering: The role of STEM interventions. *Journal of College Student Development, 57*(6), 742–747.

Romasanta, L. R. (2016). *Students as experts: Using photo-elicitation facilitation groups to understand the resiliency of Latina low-income first-generation college students.* Retrieved from ProQuest.

Roy, J. (2018). *Engineering by the numbers.* American Society of Engineering Educators. Retrieved from www.asee.org/papers-and-publications/publications/college-profiles

Saldaña, J. (2016). *The coding manual for qualitative researchers.* Thousand Oaks, CA: Sage.

Samuelson, C. C., & Litzler, E. (2016). Community cultural wealth: An assets-based approach to persistence of engineering Students of Color. *Journal of Engineering Education, 105*(1), 93–117.

Smith, W. A., Yosso, T. J., & Solórzano, D. G. (2011). Challenging racial battle fatigue on historically White campuses: A critical race examination of race-related stress. In R. D. Coats (Ed.), *Covert Racism* (pp. 211–237). Boston, MA: Brill.

Society of Women Engineers (2018). *SWE fast facts infographic.* Retrieved from https://research.swe.org/

Solórzano, D. G., Villalpando, O., & Oseguera, L. (2005). Educational inequities and Latina/o undergraduate students in the United States: A critical race analysis of their educational progress. *Journal of Hispanic Higher Education, 4*(3), 272–294.

Solórzano, D. G., & Yosso, T. J. (2002). Critical race methodology: Counter-storytelling as an analytical framework for education research. *Qualitative Inquiry, 8*(1), 23–44.

Stevenson, A. D., Gallard Martínez, A. J., Brkick, K. L., Flores, B. B., Claeys, L., & Pitts, W. (2019). Latinas' heritage language as a source of resiliency: Impact on academic achievement in STEM fields. *Cultural Studies of Science Education, 14*(1), 1–13.

SU Charter (2019). Retrieved from www.asu.edu/about/charter-mission-and-values

Valencia, R. (2010). *Dismantling contemporary deficit thinking.* New York, NY: Routledge.

Yosso, T. J. (2005). Whose culture has capital? A critical race theory discussion of community cultural wealth. *Race Ethnicity and Education, 8*(1), 69–91.

Yosso, T. J. (2006). *Critical race counterstories along the Chicana/Chicano educational pipeline.* New York, NY: Routledge.

Yosso, T. J., & Burciaga, R. (2016). *Reclaiming our histories, Recovering community cultural wealth.* Los Angeles, CA: Center for Critical Race Studies at UCLA.

Yosso, T. J., Smith, W. A., Ceja, M., & Solórzano, D. G. (2009). Critical race theory, racial microaggressions, and campus racial climate for Latina/o undergraduates. *Harvard Educational Review, 79*(4), 659–690.

# 8 Latina Undergraduates in Engineering/Computer Science on the US–Mexico Border

## Identity, Social Capital, and Persistence

*Erika Mein, Helena Muciño Guerra, and Lidia Herrera-Rocha*

In spite of the growing Latinx population in the United States, the number of Latinxs—particularly Latinas—pursuing careers in STEM fields is disproportionately low. While the number of Latinas in engineering and computer science majors has increased over the last two decades, these numbers have not necessarily translated into persistence to graduation and into the profession (Aguirre-Covarrubias, Arellano, & Espinoza, 2015; Forrest, Bennett, & Chen, 2018). According to the National Science Foundation (2017), Hispanic women comprised just 2% of the science and engineering workforce; similar numbers can be found for Hispanic women in the computing workforce (National Center for Women and Information Technology, 2019).

Research since the 1970s has highlighted the "double bind" that minoritized women in STEM face (S. M. Malcolm, Hall, & Brown, 1976), which refers to the interlocking barriers tied to both racism and sexism that can negatively impact minoritized women's (in this case, Latinas') entry into and persistence in engineering. More recent iterations of "double bind" theory have highlighted the different texture of challenges that minoritized women in STEM experience in contemporary times; as L. E. Malcolm and Malcolm (2011) argue:

> Now it is less about rights versus wrongs and more about support versus neglect; less about the behavior of individuals and a culture that was accepting of bias as "the natural order of things" and more about the responsibilities and action (or inaction) of institutions.
>
> (p. 163)

This double bind can be exacerbated by sociopolitical factors—such as access to immigrant visas—that create additional barriers to persistence for minoritized women.

In the face of the double bind, some of the social and institutional barriers faced by Latinas in engineering may be mitigated by attending a Hispanic-serving Institution (HSI) (Camacho & Lord, 2011). One reason for increased Latinx student success in engineering at HSIs can be attributed to the greater presence of Latinx mentors and role models at HSIs: "Mentors guide student success, provide a template for professionalization in the field, and can be influential in leading students to prestigious positions in industry or to pursue higher degrees." (Camacho & Lord, 2011, p. 142)

This study examines Latina student success in engineering and computer science by focusing on the social resources that contribute to Latina identities and persistence in Mechanical Engineering and Computer Science at one HSI on the US–Mexico border. Persistence is defined both as completion of the major to graduation, and continuation to employment in the profession or pursuit of graduate studies in the field. One key element of persistence is resilience, defined as a "dynamic and multidimensional process through which individuals experience positive outcomes despite exposure to significant adversity" (Kuperminc, Wilkins, Roche, & Alvarez-Jimenez, 2009, p. 214). Using a case study approach to illuminate the experiences of four undergraduate Latinas—two in Mechanical Engineering and two in Computer Science—we demonstrate the significant role of social capital in contributing to Latinas' persistence in the major and into the field.

## Theoretical Framework: Identity, Social Capital, and Academic Success

This study adopts a sociocultural perspective on identity, with a particular focus on identity as produced in and through social practices. From a sociocultural perspective, identity is tied to learning and participation in socially situated practices (Lave & Wenger, 1991). We draw in particular on Gee's (1989) notion of identity as D/discourse, that is, "saying(writing)-doing-being-valuing-believing combinations" that signal "ways of being in the world" (p. 6). The fields of mechanical engineering and computer science each include their own predominant norms of speaking and practice; to become part of the field means learning the socially accepted ways of speaking, acting, and interacting as mechanical engineers or computer scientists. In this way, one becomes recognized as a particular "kind of person" within that particular social world (Gee, 1989).

In connection with a sociocultural understanding of identity, this study also draws on the sociological concept of social capital to examine the experiences of Latinas in Mechanical Engineering/Computer Science (ME/CS) at one university. Social capital has been represented as the accumulation, distribution, and exchange of resources in and through social networks (Bourdieu, 1985; Coleman, 1988; Lin, 1999). Lin (1999) defines social capital as "resources embedded in a social structure which are accessed and/or mobilized in purposive actions" (p. 35). In this way, the notion of social resources, within the social capital framework proposed by Lin, lies at the intersection of structure and agency and "contain[s] three elements: the structural (embeddedness), opportunity (accessibility), and action-oriented (use) aspects" (p. 35). Along with other social capital theorists, Lin posits that greater access to and accumulation of social resources produces better social and economic outcomes for actors.

In the higher-education literature, researchers have shown a positive relationship between social capital, sense of belonging, and academic success among Latinx students (Maestas, Vaquera, & Muñoz Zehr, 2007; Nuñez, 2009). Sense of belonging, here, is understood as the positive interactions and attachments experienced by an individual (Baumeister & Leary, 1995) and fostered within institutional contexts (Reay, David, & Ball, 2001). Within engineering education, scholars have analyzed social capital in a variety of ways. In one study of sophomore-level engineering students at two public, predominantly White institutions, students emphasized the importance of faculty, and access to faculty, in their academic success as engineering majors (Brown, Street, & Martin, 2014). In another study that focused on first-generation female Hispanic engineering majors at a large, diverse public university, researchers used a case study approach and found that school personnel served as a significant source of social capital that influenced students' decisions to pursue engineering (Martin, Simmons, & Yu, 2013). Our study builds on these previous findings to examine the role of social capital and social resources in the identities and experiences of Latinas who persist through undergraduate ME/CS into graduate school or the profession.

## Research Context and Methodology

The findings presented in this chapter stem from a larger, NSF-funded ethnographic study of Latinx students' experiences in senior-level capstone ME/CS courses and entry into the profession. The study took place at a large, public, research-intensive HSI located on the US–Mexico border. At the time of this study, university enrollment included more than 25,000 students, 80.2% of whom were Hispanic/Latinx, an additional 4.1% being Mexican nationals. Of this student body, 73% were Pell eligible, and 50.9% were first-generation college students.

Given the focus on students' experiences within the social context of ME/CS, our chosen methodological approach for the larger study was ethnographic in orientation. Ethnography, which involves a combination of observational, descriptive, and interpretive methods (Hammersley & Atkinson, 2007), is used in a variety of disciplines as a way to investigate complex sociocultural phenomena with the understanding that the human experience is irreducible. For the purposes of this chapter, we employ a case study methodology (Yin, 2018) in order to illuminate the experiences, and patterns across experiences, of four focal participants: two Latina Mechanical

Engineering graduates, and two Latina Computer Science graduates. In constructing these case studies out of our larger body of data, we relied on three primary ethnographic data sources: (1) participant observation and field notes; (2) in-depth interviews; and (3) artifacts. These data, along with video recordings and questionnaires, were collected over the course of two academic years as part of our larger study.

The larger study focused on 27 students enrolled in senior capstone courses in ME/CS: 17 men and 10 women. The research team, which was comprised of two education faculty and two doctoral students in education (three out of four researchers identified as Mexican or Latinx), followed six teams of students enrolled in ME/CS senior capstone courses. Each semester, researchers on our team conducted more than 30 hours of participant observation in senior-level ME/CS capstone courses—one semester for ME and a two-semester sequence in CS—with students on selected focal teams. Those focal students then participated in a series of three in-depth interviews, which were conducted using the Seidman (2013) three-part interview method. The first two interviews took place one week apart at the end of the first semester or year of participant observation, while the third and final interview took place within 8–12 months after graduation, to learn more about students' experiences moving into the profession (or not). Interviews were conducted in both Spanish and English, depending on the preference of the participant. Finally, the research team collected artifacts, including course syllabi and course projects, to gain a more comprehensive understanding of the context of the senior capstone courses and course projects.

Data analysis of the interviews and field notes was an iterative process, involving multiple stages. The focus of our analysis was the factors that shaped our participants' experiences of persistence in engineering—particularly for our Latina participants. We focused primarily on the interview data for the first two in-depth interviews, which were cross-checked with the participant observation and artifact data. All interviews were coded in NVivo using an "open coding" and "focused coding" approach (Emerson, Fretz, & Shaw, 2011). Once initial themes were established, we returned to the theoretical literature, where the concept of identities (Gee, 1989) and social capital/social resources (Lin, 1999) gained particular relevance in relation to what we found in the initial round of coding.

With identity and social capital as a frame, we initiated a second round of more focused coding, and focused on four of the ten participants. These participants were selected based on the representativeness of both their experiences and their fields (two in mechanical engineering and two in computer science). In our subsequent rounds of coding, we focused on the social resources identified by participants, which in turn shaped the development of each participant's case. In the case studies presented, we attempt to provide a holistic representation of each participant's experience based on their own narratives and our participant observation. In the next section, we present each of the four cases. We start with Alicia and Daena, two Mechanical Engineering students, and then present the cases of Alejandra and Andrea, both in Computer Science (all names are pseudonyms).

## Case Studies: Latinas in Mechanical Engineering and Computer Science

### Case One: Alicia

Alicia was a Mechanical Engineering major who graduated from the program during the course of the study. She grew up on the Mexican side of the border, where she did her K–12 schooling in private schools in Ciudad Juarez. Although she moved to the United States after she was accepted into college, she crossed the border on a weekly basis to visit her family. Her father worked in the medical field in Mexico, which made her consider going to medical school for several years. Her mother was an engineer who gave up her career to take care of Alicia and Alicia's little sister; at the time of the study, her mother assisted her father at the private clinic they had in Juarez. Alicia reported that her family was always supportive of her and her academic decisions. In terms of language experiences, Alicia expressed feeling more comfortable using Spanish to communicate during teamwork interactions. She said that only once during her undergraduate studies did she have to use English to communicate with her team, and although her first language was not English, she reported being able to understand and communicate effectively.

One of Alicia's reported passions was going to the gym, so after being admitted to college she started taking boxing classes, where she met another female student who was about to graduate. That student recommended the Mechanical Engineering program to Alicia. After that conversation, Alicia declared ME as her major and started her pathway in engineering.

For Alicia, having friends, especially female friends, was an important source of academic and emotional support during her undergraduate studies. However, after her close friend Julia graduated a semester before her own scheduled graduation, she felt alone and isolated:

> El semestre pasado ya me tocó sin ella y sentí como si acababa de entrar a la escuela. Entonces, sí, este semestre sí ando como que muy solita, sin amigos, no conocía a nadie en las clases, porque todos se habían graduado de que antes o así. ["Last semester I was without her and I felt like I just had entered school. Then, this semester, I am very lonely, without friends, I don't know anyone in the classes, because all have graduated or so."]
>
> (author's own translation)

In one interview, she expressed aloud that maybe engineering was not for women after all, based on her own instances of feeling relegated to the sidelines. One instance took place after an episode involving a final project with her team, where her teammates requested that she change the fonts on the written product and that she work to make the presentation more aesthetically appealing, but she did not have the opportunity to contribute in a substantive way to the content of the final project. In one interview, she summarized this perception from male counterparts in the following way:

> Sí sentíamos mucho de que hasta los mismos niños con los que platicábamos era como que, 'Ay,' o sea, como que, o sea, para mí, graduarme es como un súper logro porque como que siento que todos ellos pensaban que no, como que: 'Ay, Alicia no va a poder,' 'Ay, Alicia no sé qué.' . . . como que nos trataban así como que: 'Ay, ellas sí están tontas'. . . . ["We felt very much that even the boys with whom we talked were like 'Ay,' . . . it was like . . . for me, graduating was a super achievement because I feel that all of them were thinking, like, 'Ay, Alicia is not going to make it', 'Ay, Alicia, I don't know' . . . like they treated us like . . . 'They [girls] are dumb'. . . ."].
>
> (author's own translation)

During her second interview, which took place one month prior to her graduation from ME, Alicia expressed not having a clear idea of what she would do after graduation. She was on the fence about whether to pursue a master's in engineering or to go into the business world by opening her own yoga studio in Juarez. She expressed that she had her father's financial support for either option. After graduation, she ended up doing a travel tour through Asia for six weeks with friends. After that experience, another friend contacted her to offer her a job at a company outside of the field of engineering. She worked at that company for a couple of months before starting her master's in Civil/Environmental Engineering.

During her final interview one year after graduating with her bachelor's, Alicia expressed feeling happy to be back in school, as she missed the routine of going to classes and to the gym. She was also happy that her new program was more dynamic, hands-on, and required more time in the field than her undergraduate program. In her final interview, she shared her plans to apply to an internship the following year, and she expressed wanting to continue living on the border because being close to her family was important to her.

### Case Two: Daena

Daena was a Mechanical Engineering major who graduated during the study. She completed elementary and early middle school in Chihuahua,

Mexico, and finished her K–12 education in the United States. Her father worked as a mechanic in the construction field and did not complete post-secondary studies. Her mother graduated from nursing in Mexico but gave up her career to take care of Daena and her siblings. In terms of language, Daena expressed experiencing some struggle with her transition from speaking Spanish at school and home to being taught in and using English to communicate with classmates upon moving to the United States. She was observed using border varieties of both languages to communicate with classmates and friends but only English with professors or people with higher status.

From the time she was a child, Daena observed her father fixing cars, and she grew up with an interest in mechanics; she knew from a young age that this would be her professional pathway in the future. Because of her interest in mechanics and as a result of her sociability and outgoing nature, she got involved in mechatronics, robotics, and other team projects related to the engineering field during her high school years. These experiences led to opportunities to travel and participate in national competitions, which led to networking experiences during high school, particularly with one aeronautics company. She began her undergraduate studies at a different university from the one where this study took place, where she struggled unsuccessfully to gain admission into the engineering program. Given her struggle with the courses and requirements, she ultimately decided to transfer to the university where this study took place.

Daena expressed that having friendships was very important to her, particularly throughout her experiences of struggle and persistence to obtain her ME degree. She also shared that some of her previous friendships had turned into professional networking relationships. One in particular led to a job offer, which she ultimately took. She summed up her view of social capital in her career, stating that "people skills are very important," especially in engineering.

Another experience that was important for Daena was being part of organizations, sororities, conferences, and public events related to the ME field. She was particularly drawn to being part of Hispanic engineering and Hispanic women engineering organizations. This was related to her clear expression of identity as being a "proud Latina," which she demonstrated in a variety of ways. One of the clearest examples of this expression was in her founding and establishment of a female engineering student organization, which took place when she transferred to the university. Although she expressed being accustomed to being surrounded by men, her closest friends were "Hispanic females," as she called them.

During her final semester in the ME program, she was interviewed by several companies. Before graduating, she had a job offer from an aeronautics company. The job offer stemmed from her friend-turned-professional-contact, who served as the CEO of the company.

## Case Three: Andrea

Andrea was a Computer Science major who graduated during the study. She was raised on the US side of the United States–Mexico border, and completed all of her K–12 schooling in the United States. Her parents were 15 and 16 years old when they had her, and separated just months after her birth. She was raised by her mother and stepfather, neither of whom attended college. While growing up, she was really interested in videogames, a pastime she shared with her father, who had completed an online bachelor's degree. In school, she was placed in the Gifted and Talented Education (GATE) program from a young age. During high school, she earned good grades; she took college-level dual-credit coursework and demonstrated a strong interest in STEM-related courses.

Although Andrea was interested in CS during high school, she hesitated to follow that pathway. But one of her uncles, who was studying computer science at the university, encouraged her to pursue CS at the same school and offered to support her with her courses and assignments during her first year of college before he graduated. Despite being told that she was good at math and would thus make it in CS, Andrea was fearful of pursuing this major, which she attributed to persistent messages that she heard during high school about her not being ready for CS, as it was something that one needed to prepare for from an early age.

In spite of these hesitations, Andrea ultimately applied for admission to the CS program and was accepted. During her undergraduate studies, she received a scholarship to pay for her education with the support of her mentor, who was a faculty member in CS and director of the scholarship program. Her scholarship required her to take a fast track by taking dual— undergraduate and graduate—credits and to commit to an internship with the federal government.

One important learning experience for Andrea during her undergraduate studies was participating in professional conferences, including the Women's Cybersecurity Conference three years in a row with support from her mentor and financial support from the college. Participating in conferences gave her the opportunity to be exposed to and get involved in specialized areas of the field. She highlighted the impact of participation in conferences, saying, "it's a lot of networking where you make connections and I still keep in contact with some people. And they give you mentorship or revise your resume. . . ." She also commented that being surrounded by females in these conferences made her feel confident in her program of study and her own knowledge base.

Andrea also emphasized the importance of peers and friends who supported her academically, emotionally, and professionally. For example, a big sister in her sorority helped her complete the application for a government internship doing cybersecurity at a prestigious university-affiliated research center during her second year of college. In that experience, she met her supervisor, whom she considered her mentor. Her mentor supported her learning and gave her the opportunity to advance her knowledge and skills through participation in conferences and other activities.

She expressed love for her hometown and university, especially because she had the opportunity to be close to her professors given the size of the university. Upon graduation, Andrea obtained another internship at the same research center where she had previously interned. She planned to continue with her master's at the same college after finishing her internship.

### Case Four: Alejandra

Alejandra was a Computer Science major who graduated from the program during the study. She completed elementary and middle school in Mexico and high school in the United States while living in Mexico. Her mother was a medical assistant who had wanted to pursue a career in medicine but was not able, given financial constraints in her family. She did not grow up with a father, but she had an uncle who was a father figure and who would always tell her she was smart. During her time in high school and college, she traveled across the border every day to attend school. In her interviews, she described crossing the border by foot on a daily basis, suffering all kinds of weather conditions, suspicious activities, and sexual harassment if she crossed the border going back to Mexico late at night wearing a skirt. Her *transfronteriza* or border-crossing experience was a key aspect of her undergraduate experience. Given her cross-border experiences, she was fully bilingual, speaking Spanish at home and both English and Spanish at school.

Alejandra described always liking to be around technology while growing up. She got her first computer when she was 6 years old, which she used to play STEM-like adventure games, instead of playing with "girl toys," due to her uncle's influence. She said that she always knew she wanted to pursue a STEM-related major. During her senior year of high school, she received a full-tuition scholarship to study engineering at the university where this study took place. As part of the scholarship, the university organized a special event, where she met a female Latina professor in the CS program who eventually became her mentor. Her mentor often talked to her about the need to have more women in a male-dominated field, which, she said, would help make the field more inclusive. During their initial encounter, Alejandra shared her concern about not knowing how to code but received comfort and encouragement to enter the field.

In her interviews, Alejandra demonstrated an awareness about being a minority, not only because of her gender, ethnicity, and citizenship but also because of her intersecting identities as a Latina in CS. One comment highlighted the contradictions she felt: "Being a Latina is something beautiful, although it sucks sometimes." She shared that because she was a Mexican national studying on a student visa, she struggled with internship interviews, despite her work experience and high GPA. In one instance, she was called for an interview, only to be told that "they don't take international students for entry-level positions or for internships." This experience left her heart-broken and in tears.

Alejandra also expressed being happy to be part of the university and border community, with both its advantages and disadvantages. For example, she was fearful of the immigration climate, which had made her feel unwanted and unwelcome. She was concerned about her future academic and professional opportunities after graduating from her master's degree program, which she began upon completion of her undergraduate studies. On the whole, Alejandra emphasized the importance of her family, friends, and boyfriend as her primary support to continue on her CS pathway.

## Discussion

Each of the four Latinas presented in these cases had a unique trajectory into and through ME/CS studies. All of the participants expressed different kinds and degrees of social capital, which shaped their identities and positively impacted their resilience and persistence. In this section, we will discuss each case individually and then highlight patterns across the four cases.

### Precarious Ties to and in Engineering

Of the four cases presented, Alicia had the weakest connection to her major, mechanical engineering. While she persisted successfully through her undergraduate ME program and ultimately entered a master's program in engineering, there were moments during her senior year when her persistence into the profession seemed precarious. One such instance occurred in her second interview, when she commented that she was considering opening a yoga studio in Mexico upon graduation. Her precarious connection to ME seemed to stem in part from the reported sense of isolation that she experienced during her senior year and the sense that she did not belong in engineering, which was reinforced by certain interactions with her team in her senior design project. In one interaction, she was asked to make revisions to a team project by making the presentation "look prettier" but not by making substantive contributions to content. This interaction made her question whether she fit, particularly as a woman, in engineering. Alicia's ties to her senior design team could be categorized as weak social ties, which ultimately detracted from, rather than supported, her persistence in engineering.

These weak social ties were offset by two other social resources: her family and her friends in the major. Alicia came from a middle-class professional family in Mexico that she perceived as supporting her (both emotionally and financially) in whatever decision she made upon graduating—that is, whether to continue in engineering or to start a yoga studio. A second source of social capital for Alicia were her friendships, which she said provided social and emotional support during her studies. When all of her friends graduated before her, midway through her senior year, Alicia felt isolated and questioned whether she belonged in the major, though she demonstrated resilience in sticking with the program through graduation. As a noted contrast, in the third interview, which took place nearly a year after graduation, Alicia had started a graduate program in engineering, which meant that she continued her trajectory in the field. In her master's program, she reported feeling a greater sense of belonging based on both the relationships that she was forming and the hands-on approach to learning.

Two notable forms of social capital were absent from Alicia's account of her undergraduate experience: teachers and mentors, on the one hand, and academic/professional peers, on the other. Alicia mentioned neither of these in her interviews, and they were not apparent in our observations.

### Strong Peer Networks in Engineering

The second case represents a contrasting case to Alicia. Daena was also a mechanical engineering major, and she reported feeling a strong connection to the field from when she was a young child and learned to fix cars alongside her father, who was an auto mechanic and construction worker. Her early affinity for mechanics led her to get involved in experiences in high school that reflected these interests, and it was through these experiences that she gained early exposure to travel, participation in national competitions, and networking with peers in the profession. From an early point in her academic trajectory, Daena began accumulating social resources that positively contributed to her belonging and persistence in engineering. Even though she experienced a major obstacle early in her undergraduate studies, when she struggled to gain admissions to an engineering program at a predominantly White institution, she succeeded in overcoming that struggle by transferring to another university, an HSI in her hometown.

Upon transfer to the institution where this study took place, Daena thrived. She started a student organization focused on women in engineering, and she cultivated a wide network of friends and peers in engineering. The social capital that she developed at the HSI contributed to her sense of resilience and belonging in the major; moreover, it had a direct impact on her persistence into the profession when one of her engineering peers connected her to the aeronautics company where she ultimately accepted a job offer.

While Daena boasted a strong network of peers and personal friends in her major, which had a significant impact on her successful movement into the profession, she did not place emphasis on another form of social capital in her undergraduate studies: teachers and mentors. In this way, Daena's experience was not unlike that of Alicia, who was also a mechanical engineering major. The point of contrast between their two experiences, however, can be found in the peer networks that Daena cultivated. While Alicia's social ties in engineering were personal in nature, Daena built both personal and professional peer networks. It was through professional networks that Daena received her first step into the profession through the aeronautics job offer.

### Multiple Social Resources Supporting Persistence in CS

The third case presented, Andrea, helps shed light on how multiple types of social resources can be leveraged in support of persistence and success in undergraduate computer science. Despite being positioned as smart from a young age, Andrea showed hesitation in pursuing computer science as a major because she felt she had not gotten involved in the field early enough to be successful. What aided Andrea in overcoming these initial doubts was the support of a family member, her uncle, who was a computer science major and who encouraged her to apply to the program at the same university. Her uncle served as an important social resource in her decision to enter the program. At the university, Andrea developed a broad network of friends and peers, one of whom encouraged her to apply for an internship in cybersecurity. Andrea received the internship, which in turn led her to develop a relationship with her primary mentor in computer science, who was a faculty member in the program. The encouragement of her sorority sister to apply for the internship represented an important social resource that helped shape Andrea's pathway.

The relationship with her mentor in computer science, which emerged as a result of the cybersecurity internship, became another significant social resource. Her mentor connected her with both scholarship support and professional opportunities, such as participation in a yearly cybersecurity conference for women. In this way, the connection with her mentor led directly to material outcomes, in terms of financial support in the program, in the opportunity to engage in professional learning at conferences, and in exposing her to a supportive (national) network for women in computer science. This national network, in turn, helped affirm her own decision to go into computer science, contributing to the development of her computer scientist identity and her resilience throughout the program. The internship in which she participated ultimately led to a full-time paid internship post-graduation, which assured her persistence in computer science not only through graduation but into the profession.

### Sociopolitical Factors Constraining Persistence

The fourth and final case presented, Alejandra, illuminates the complexity of social resources within particular sociopolitical contexts, where greater social capital does not necessarily confer material opportunities or outcomes. Alejandra grew up in a lower middle-class family on the Mexican side of the border, where her single mother supported her financially and emotionally as she completed high school and entered university on the US side of the border. Her affinity toward computer science started at a young age and was nurtured by an uncle who served as a father figure and who told her that she was smart. Both her mother and uncle represented important social resources that initially shaped her pathway into and resilience in computer science.

At the university, Alejandra received a full scholarship to pursue an engineering major. As part of the scholarship program, she came into contact with a Latina professor of computer science. This professor became an important social resource for Alejandra, not only in encouraging her to go into the field despite her self-doubt but also by showing her that it was possible for a Latina to be a computer scientist, which was critically important for her own identity development.

While Alejandra had multiple social resources, both familial and academic, that assisted her in her undergraduate pathway, these resources proved insufficient in pursuing next steps in computer science after completing her master's degree. As a daily border-crosser (*transfronteriza*) residing in Mexico and studying in the United States, Alejandra was able to secure a student visa. Having a student visa opened up doors for studying, but it did not pave the way for her to secure full-time employment in the United States; in fact, she felt mocked during job interviews because she did not have the correct visa to be able to work. While Alejandra completed her studies in computer science and persisted into a master's program, her persistence into the profession was still unclear at the time of this study, given the constraints imposed by her visa situation.

## Conclusion

These four case studies elucidate ways in which Latinas' social capital contributed to their identity development, and ultimately their resilience and persistence, in ME/CS undergraduate programs and into ME/CS professions or graduate programs. At least four types of social resources were visible across one or more cases: family resources; friendship resources; peer resources; and mentor resources. The first two types—family and friends—were identified across all four cases. Familial and peer social resources played the role of social and emotional support for all four women during their studies and provided Andrea and Alejandra encouragement to pursue engineering. Peer resources were identified in three of the four cases and played a particularly important role in one case (Daena), where a peer/professional contact led to a job offer. In another case, that of Alicia, there was not a clear identification of peer resources in her major, and Alicia felt the weakest connection to her program (Mechanical Engineering) during her undergraduate studies (though she ultimately persisted into graduate school in engineering). Finally, the fourth type of resources—mentor resources—were visible across two of the four cases (Andrea and Alejandra). Mentor resources were particularly important for connecting participants to financial and professional opportunities, including scholarships and access to national conferences. That the two students who identified mentor resources were in computer science is significant, as this program was explicit about its mentoring of students into the profession, and the senior capstone course in CS was structured in a way that reflected this explicit approach.

In addition to the social resources identified by the Latinas in our study, another finding related to the limitations of social capital. While all four participants signaled clear benefits from the social resources that they identified, one of the four (Alejandra), struggled to persist into the profession because she did not have access to a work visa at the time of the study. Her case helps illustrate the ways in which social capital can be materially beneficial (with access to scholarships and conference opportunities) but insufficient under particular sociopolitical conditions—in this case, for an international student without a work visa who was seeking employment.

Taking into consideration the limitations of social capital outlined previously, the findings from this study still point to a significant role that social capital plays in supporting undergraduate Latinas' persistence through engineering studies and into the profession. This study has two clear implications for undergraduate programs: one connected to mentoring and the other connected to peer resources. The differential access to mentors found among participants in the two focal programs studied—mechanical engineering and computer science—sheds light on the ways in which programs and faculty can intentionally provide mentor resources for Latina students.

## Recommendations for Practice

Our findings have several significant implications for practice for institutions to support the identity development, resilience, and academic success of Latinas in engineering. First, the findings illuminate the critical importance of mentoring in the experiences of Latina engineering students. In the case of computer science, mentoring was built into the structure of the required senior capstone course; this model can serve as an example for other programs. Mentoring by Latina faculty members, in particular, can positively impact the trajectory of Latinas in ME/CS, as seen in the case of Alejandra. In this way, increasing the number of—and providing sufficient support for—Latina faculty in engineering and computer science can lead to both greater numbers and greater success of Latinas who choose to pursue the major and who persist through graduation and into the profession.

In addition to mentor resources, undergraduate programs can be intentional about fostering the development of peer resources among students, both within the classroom—through pedagogies that support peer learning and through the intentional creation of study groups—and outside of the classroom, through the explicit support of student organizations that help build a sense of community and contribute to the expansion of peer networks among Latinx students. The strategic actions taken by institutions to support Latinx resilience, persistence, and success in engineering and computer science would, in turn, help contribute to equity in STEM in higher education and beyond.

## Acknowledgement

The material presented in this chapter is based on work supported by the National Science Foundation, Award #1734967. All findings, conclusions, and recommendations are those of the authors and do not necessarily reflect the views of the National Science Foundation.

## References

Aguirre-Covarrubias, S., Arellano, E., & Espinoza, P. (2015). "A pesar de todo" (Despite everything): The persistence of Latina graduate engineering students at a Hispanic-serving institution. *New Directions for Higher Education, 2015*(172), 49–57. doi:10.1002/he.20152

Baumeister, R. F., & Leary, M. R. (1995). The need to belong: Desire for interpersonal attachments as a fundamental human motivation. *Psychological Bulletin, 117*(3), 497–529. doi:10.1037/0033–2909.117.3.497

Bourdieu, P. (1985). The forms of capital. In J. G. Richardson (Ed.), *Handbook of theory and research for the sociology of education* (pp. 241–258). New York, NY: Greenwood.

Brown, S., Street, D., & Martin, J. P. (2014). Engineering student social capital in an interactive learning environment. *International Journal of Engineering Education, 30*(4), 813–821.

Camacho, M. M., & Lord, S. M. (2011). *Quebrando fronteras*: Trends among Latino and Latina undergraduate engineers. *Journal of Hispanic Higher Education, 10*(2), 134–146. doi:10.1177/1538192711402354

Coleman, J. S. (1988). Social capital in the creation of human capital. *American Journal of Sociology, 94*, S95—S120. doi:10.1086/228943

Emerson, R. M., Fretz, R. I., & Shaw, L. L. (2011). *Writing ethnographic fieldnotes* (2nd ed.). Chicago, IL: University of Chicago Press.

Forrest, E. C., Bennett, C. T., & Chen, X. (2018). *First-generation students: College access, persistence, and postbachelor's outcomes: Stats in brief* (NCES Publication No. 2018–421). Washington, DC: National Center for Education Statistics.

Gee, J. P. (1989). Literacy, discourse, and linguistics: Introduction. *Journal of Education, 171*(1), 5–17. doi:10.1177/002205748917100101

Hammersley, M., & Atkinson, P. (2007). *Ethnography: Principles in practice* (3rd ed.). New York, NY: Routledge.

Kuperminc, G. P., Wilkins, N. J., Roche, C., & Alvarez-Jimenez, A. (2009). Risk, resilience, and positive development among Latino youth. In F. A. Villlaruel, G. Carlo, J. M. Grau, M. Azmitia, N. J. Cabrera, & T. J. Chahin (Eds.), *Handbook of U.S. Latino psychology: Developmental and community-based perspectives* (pp. 213–233). Thousand Oaks, CA: Sage.

Lave, J., & Wenger, E. (1991). *Situated learning: Legitimate peripheral participation*. Cambridge, MA: Cambridge University Press.

Lin, N. (1999). Building a network theory of social capital. *Connections, 22*(1), 28–51. doi:10.4324/9781315129457-1

Maestas, R., Vaquera, G., & Muñoz Zehr, L. (2007). Factors impacting sense of belonging at a Hispanic-serving institution. *Journal of Hispanic Higher Education, 6*(3), 237–256. doi:10.1177/1538192707302801

Malcolm, L. E., & Malcolm, S. M. (2011). The double bind: The next generation. *Harvard Educational Review, 81*(2), 162–171. doi:10.17763/haer.81.2.a84201x508406327

Malcolm, S. M., Hall, P. Q., & Brown, J. W. (1976). *The double bind: The price of being a minority woman in science* (AAAS Publication No. 76-R-3). Washington, DC: American Association for the Advancement of Science.

Martin, J. P., Simmons, D. R., & Yu, S. L. (2013). The role of social capital in the experiences of Hispanic women engineering majors. *Journal of Engineering Education, 102*(2), 227–243. doi:10.1002/jee.20010

National Center for Women and Information Technology (2019). *By the numbers.* Retrieved from www.ncwit.org/resources/numbers

National Science Foundation (2017). *Women, minorities, and persons with disabilities in science and engineering: Occupation.* Retrieved from www.nsf.gov/statistics/2017/nsf17310/digest/occupation/overall.cfm

Nuñez, A. (2009). Latino students' transitions to college: A social and intercultural perspective. *Harvard Educational Review, 79*(1), 22–48. doi:10.17763/haer.79.1. wh7164658k33w477

Reay, D., David, M., & Ball, S. (2001). "Making a difference?": Institutional habituses and higher education choice. *Sociological Research Online, 5*(4), 1–12. doi:10.5153/sro.548

Seidman, I. (2013). *Interviewing as qualitative research: A guide for researchers in education and the social sciences* (4th ed.). New York, NY: Teachers College Press.

Yin, R. K. (2018). *Case study research and application: Design and methods* (6th ed.). Thousand Oaks, CA: Sage.

# 9 "I Learned How to Divide at 25"

## A Counter-Narrative of How One Latina's Agency and Resilience Led Her Toward an Engineering Pathway

*Dina Verdín*

Disciplinary role identities are dynamic, malleable, and are influenced by situations and interactions (i.e., sociohistorical struggles, environmental structures). Yet, students have the agency to adopt identities that they see as congruent with who they want to become. I used the lens of disciplinary role identity—which situates identity development through an interplay between disciplinary interest, beliefs about performing well and understanding content material and being recognized by others as a STEM type of person, and subsequently accepting and internalizing that recognition (Carlone & Johnson, 2007; Gee, 2001; Godwin, Potvin, Hazari, & Lock, 2016; Hazari, Sonnert, Sadler, & Shanahan, 2010; Johnson, Brown, Carlone, & Cuevas, 2011; Verdín, Godwin, & Ross, 2018; Verdín, Godwin, Kirn, Benson, & Potvin, 2019). Likewise, interest, recognition, and beliefs about one's performance (i.e., disciplinary role identity development) are fostered through activities, social practice, and agentic capabilities (Holland, Lachicotte, Skinner, & Cain, 1998). Enacting agency involves a reflective practice, and it requires that students continually negotiate their identities-in-practice as they move toward seeing themselves as math, physics, and ultimately engineering type of people. Authoring a disciplinary role identity develops through a continuous process that takes a great deal of negotiation and agentic capabilities (Holland et al., 1998).

Becoming the type of person one envisions for herself (i.e., an engineer) is an agentic act requiring intentional choices, reflectivity on the outcome of these choices, and resistance against environmental constraints. Agentic acts position individuals as active participants of their lives and allow students to explore, maneuver, and impact their environment for the achievement of a goal or set of goals (Bandura, 2001). Likewise, agentic behaviors can be manifested as resilient acts that help students position themselves as active contributors to their career trajectories (Pruyn, 1999; Sapp, Kiyama, & Dache-Gerbino, 2015). Emphasizing the agentic capabilities students hold is a way of framing their experiences and pathways through an asset-based perspective.

Students display agency when constructing their identity and when resisting oppressive social and environmental structures. Constructing disciplinary role identities is a process that occurs within contextual constraints (Schwartz, Luyckx, & Vignoles, 2011); these contextual constraints can be conceptualized as three types of environmental factors, that is, environments that are *imposed, selected,* or *constructed* (Bandura, 2001). An imposed environment may include daily situations or circumstances that a student interacts with (e.g., cultural worlds, communities of practice, school settings, microaggressions, racism, etc.). Imposed environments are created through the messages circulating about who gets to participate in STEM fields or who is recognized as a legitimate member of a community of practice. However, students do have the ability to interpret and react to their imposed environment. Even within imposed environments, students can select their environment based on their reactions and resistance. Lastly, a constructed environment requires students to actively engage in and with their surroundings; through the process of engagement, students can acquire new knowledge, dispositions, and behaviors. Each form of environment requires different levels of agency, and a student can shape their environment through the enactment of their agency and resilience. The environments that students navigate have identity-shaping consequences because identities are dynamically constructed through experiences (Oyserman, Elmore, & Smith, 2012).

This study uses the lens of agency, environmental factors, and disciplinary role identities to understand how one student came to see herself as someone who can do engineering (i.e., identity development). Kitatoi[1] is the oldest of six siblings, all born and raised in Southern California. Her parents immigrated from Mexico in the early 1980s, settling first in East Los Angeles and later making their way inland. Like many Mexican families in the Southwest, Kitatoi's parents did not speak English. Her family was of low socioeconomic status, and she would one day be the first in her family to attend (and drop out) of college. Kitatoi, while being born and raised in California, was predominately literate in the Spanish language she spoke at home. Now, in her late 30s and a mother of three, Kitatoi takes me through her journey from her remembered educational experiences, her six-year journey through community college, and how she came to see herself as someone who can do engineering.

## Description of Study

I used critical race methodology, which uses a counter-narrative approach as a "method of telling the stories of those people whose experiences are not often told" and as a "tool for exposing, analyzing, and challenging the majoritarian stories" (Solórzano & Yosso, 2002, p. 32). Counterstories bring patterns of racialized inequalities to the forefront by recounting experiences of racism both individually, by the participant, and shared experiences from scholarly work (Yosso, 2013). I used a counter-narrative and narrative approach to reveal how Kitatoi came to see herself as someone who can do engineering. A narrative approach allowed me to organize the experiences of Kitatoi in a temporally meaningful manner integrating her past, present, and future, and this narrative gives a "sense of continuity necessary for identity formation" (Rossiter, 1999, p. 62). Kitatoi's narrative is framed around agentic (resistive) acts. Individuals who are oppressed in society are often portrayed as victims, and their agency or ability to resist oppressive structures is overlooked. Kitatoi's narrative is also the story of experiences at the margins of society; thus, her experience is a counter-narrative challenging the discourse on the types of people who get to participate in engineering.

In this chapter, I share Kitatoi's retold experiences that were significant to her academic trajectory that spanned from primary education to her transition into higher education. I start with her primary and secondary educational experiences, situating her narrative in the current educational literature because they provide a rich understanding of how and why Kitatoi felt disempowered as an adult in society and why her decision to enroll in a community college was an agentic act. Following, I retell Kitatoi's narrative at community college, capturing experiences that both enabled and constrained her agency, and experiences that supported the development of her disciplinary role identities (i.e., mathematics, physics, and engineering). I end the chapter by acknowledging that the route to engineering is open for all students, especially those who are yet to see themselves as mathematics and physics capable learners.

### Remembered Experiences, Feelings, and Stories Before Community College

Seeing oneself as a type of person who can do engineering is an identity that is shaped and reshaped through social participation and practice. However, it is not possible to view learning and participation divorced from students' backgrounds and communities to which they belong. As you soon shall see, Kitatoi's journey into engineering did not begin with playing with Legos at a young age, enrolling in summer STEM camps, or through advanced high school math/science preparatory courses—a common account among privileged students studying engineering (Cruz & Kellam, 2018; Verdín, Godwin, & Ross, 2018; Warne, Sonnert, & Sadler, 2019). Instead, her pathway into engineering was against environmental constraints and, in large part, was a result of her agency, i.e., the resistive acts and choices she made in response to and against external influences. The exercise of agency and the change that comes with this exercise is a reflective process. Before I begin to tell the story of how Kitatoi came to see herself as someone who can do engineering, I must do justice to her journey by uncovering the sociohistorical context she agentically and resiliently navigated through. I start by eliciting a reflective narrative.

*Interviewer:*    Can you tell me how you came to where you are today?
*Kitatoi:*    I pretty much was aware that I didn't have any education, and I wasn't going to be able to get a job . . . or support myself. I pretty much learned maybe ten years ago that when my kids got to the age that they were going to need help with schoolwork like math, science, English, anything, I wasn't going to be able to help them because I had such a bad experience in high school. I just had very little education and just very little knowledge about everything around me. *I decided . . . I don't want to be like that anymore.* I want to know how things work. I wanted to be able to know history, know how to write a decent essay. I wanted to be able to know basic math. . . . This was my goal, to be knowledgeable so I can help the kids with their homework.

Kitatoi's reflection of her life's course and decision to change the trajectory of her life was a powerful emancipatory act toward reclaiming her agency, toward reconstructing her environment, and her resilient desire to reshape how she saw herself (i.e., seeking identity-congruence). Kitatoi's narrative is not a story of a high school dropout; in fact, she graduated from high school, barely meeting the graduation requirements and with minimal academic preparation. In her words, "I didn't get anything out of high school, I just got the fuck out of there." Kitatoi's reflection reveals how disempowered she felt as an adult in society, how her agency had been hindered due to an education system that failed to prepare her, and how this, in turn, would affect her children.

Kitatoi's story is not uncommon. Scholars have long documented how the public educational system has shortchanged Mexican American students (Ballón, 2015; Rumberger & Rodríguez, 2010; Valencia, 1997, 2011). Kitatoi began high school in Southern California in 1997 against the backdrop of a master narrative of Mexican Americans' educational attainment, achievement, educational value (Valencia, 1997, 2004), and state initiatives prohibiting bilingual education, which shaped the narrative of how linguistic minorities were viewed in society (García, Wiese, & Cuéllar, 2011). Kitatoi was never formally enrolled in a bilingual program that would support her language development.

> In [elementary] I had a teacher in second grade who was bilingual and she helped me a lot, then I had her again in fourth grade *y otra vez* she help me a lot, then I went to fifth grade. . . . I remember crying. . . . I realized that the teacher not only didn't speak Spanish . . . [he] . . . told me that I had to deal with not having her [bilingual teacher] around to

keep helping me. *Ósea, que ya ni modo, que me aguanté.* . . . [In] middle school . . . I was just lost in understanding a lot of things, and teachers just sort of didn't pay attention to me 'til I messed up on something, then they realize it was a language thing *y nomas me decían* well, "you gotta try harder" or some shit like that. . . . Some me *regañaban* that I was falling behind or I'm at a lower level than I should be at or tell me that I needed to do something about it . . . a teacher told me that I needed to stop . . . playing around . . . with my friends so much and focus on the homework or readings. I never went out and had no friends in middle school. . . . In high school I remember I would try to cheat my way to a good grade cause I didn't know what else to do and one teacher caught me . . . she was disappointed but just said I needed to do better, and *I remember being confused like, what do you mean better? How?* But she didn't tell me.

Kitatoi recounted the teachers' relationship with her as inauthentic, lacking care and compassion for her educational struggle. Her teachers tended to be more concerned with content acquisition as opposed to Kitatoi's subjective reality. In many ways, teachers were interpreting her lack of content knowledge as a form of off-putting behavior, signifying that she didn't care. Kitatoi's lack of academic preparedness, in the eyes of her teachers, was her fault (i.e., victim blaming). With minimal support and care from her teachers throughout middle school and high school, Kitatoi resiliently attempted to act on her environment by cheating her way to a good grade. While her reconstructed environment sat in tension with the school's moral code, Kitatoi was enacting her agency to resist the imposed school environment that left her to care for herself using a resource that made sense to her because "[she] didn't know what else to do."

Valenzuela's (1999) extensive ethnographic account observed how students at one predominately ethnic minority high school were subjected to linguistic and cultural divestment where the imposed environmental structures (i.e., policies and practices) subtracted students' culture in favor of assimilation. The statewide discourse and public perception of bilingual education of students who are English-language learners were fiercely contested issues that subtracted students' culture or educational attainment and impacted how Kitatoi saw herself and who she wanted to become.

The kids [in] . . . ESL [English Second Language] classes . . . everything was the lowest level . . . lowest level English, lowest level math, lowest level. . . . It was just like, if you're part of this group and you're going to this class, you're not going to get really far in life. You're not going to go to college. . . . I just didn't want to be associated with them because of this stereotypical thing of, "All the Mexicans hang out in the back of the school . . . they're not going to get very far. . . ." because that's what the vibe was in high school.

The high school Kitatoi attended was guilty of reproducing inequalities among students by preparing certain types of students differently than others. Solorzano and Solorzano (1995), Valencia (2011), and Ballón (2008, 2015), among others, have argued that the students in these ESL classes had unequal schooling conditions (i.e., ability grouping, curriculum differentiation, and low expectations) that stunted their opportunities by offering a curriculum that only prepared them for low-paid and low-skilled jobs. And as Valencia (1997) would argue, the ESL students were receiving subtractive schooling by having resources removed (i.e., regular-tracked courses and college preparatory courses). Kitatoi's refusal to enroll in ESL classes was a resistive act, not on the part of being bilingual or Mexican American, but on the subject positioning and unequal opportunities being afforded to language-minority students in those classes. Her perception of Mexican Americans and people with English as a second language was constantly at odds with who she wanted to be, in large part due to an imposed reproduction of a deficit perspective that dominated the current discourse. The master narrative of how Mexican Americans were positioned in society was rooted in deficit and racist views intended to systematically disenfranchise a growing community in the Southwest (Solorzano & Yosso, 2000). Kitatoi recounted how she saw society's positioning of people like her.

> I never saw or knew any Mexicans with English as their second language more than like field or construction workers . . . house cleaners or housewives or working at restaurants and so I just figured that that [English as a second language] was the reason why.

Kitatoi, as an adolescent, was impressionable and vulnerable to the single narrative she saw in her surroundings, unaware that society had systemically disempowered and disenfranchised people like her (see Gonzalez, 2013; Tejeda, Martinez, & Leonardo, 2000; Valencia, 1997, 2004). Although Kitatoi enacted her agency by resisting the subject position placed on ESL and Mexican American students, she was navigating an educational environment that had low expectations and negative stereotypes of students like her. Her experience in high school was rooted in long-standing racist stereotypes about Mexican Americans as indifferent toward and devaluing education, denoting an imposed deficit narrative (Valencia, 1997; Valencia & Black, 2019). While Kitatoi agentically resisted the master narrative by constructing her own path, her action would have a long-lasting effect. Kitatoi, throughout high school, concealed the fact that she too struggled with the English language.

> I felt really behind in high school. English is my second language. I didn't really understand what I was reading. . . . I didn't really know how to read that well in English . . . how to write in English. I knew a little bit, but not where you're supposed to be at a high school

level. . . . I flunked English, freshman English, and I flunked math. I failed those classes, and I never told anybody because it was embarrassing. I wasn't understanding the teachers, I wasn't understanding the assignment. I wasn't really understanding anything. I understood very little in order to pass high school, but I really don't think I learned anything. . . . I didn't know how to write a basic essay or do basic arithmetic.

Kitatoi's feeling of disempowerment, the feeling of being disregarded and overlooked by her teachers, fractured her agency and led her to withdraw from academics. Systemic inequalities and prejudicial views of Mexican American youth, as have been discussed by scholars (Stanton-Salazar, 2001; Valencia, 2011; Valencia & Black, 2019; Valenzuela, 1999), allowed Kitatoi to enter high school with low levels of preparation. While Kitatoi graduated from high school, she "understood very little in order to pass, but I don't think I learned anything." I would argue that the imposed academic self-concept of a struggling student became overpowering.

Despite Kitatoi's depressed academic preparation, as a result of the education system that failed her, she enrolled in a local community college, ready to beat the stereotype that Mexican American youth were "not going to get very far." After graduating high school, Kitatoi was attempting to redefine herself as a capable student; however, it would seem that imposed environmental influences (i.e., shortchanged academic preparation, subtractive schooling, and societal pressures of where Mexican American women like her are positioned) would overcome her agency. Kitatoi was placed into a master narrative of the role women like her were expected to fulfill in society. After a semester, her academic trajectory took a turn, she became a full-time housewife at the age of 19:

> I decided that I was just going to get married. Wasn't going to do school, was just going to get married and have kids and that was going to be my life . . . at that time I thought [getting married] was going to be the best thing for me.

Looking backward to remembered experiences, feelings, and stories were important in moving towards understanding the sociohistorical factors that shaped Kitatoi's trajectory. Throughout her trajectory, her agency was both constrained and enabled—constrained as a result of the master narratives that imposed a certain way of being in the world and enabled when she actively resisted these narratives. Kitatoi's reflection offers a view of how agency is both enacted and suppressed by environmental factors (i.e., environments that are imposed, selected, and constructed). The dance between constrained and enabled agency ultimately shaped how Kitatoi saw herself in relation to learning and being a student. Kitatoi's identity, albeit imposed due to environmental factors, at this point in her life was that of a struggling student. In the next section, I discuss how Kitatoi's reconstructed agency allowed her to re-enroll at community college, which paved a pathway toward studying engineering and ultimately reshaped her perception of herself as a capable learner.

### Enacting Agency: Journey Through Community College

Kitatoi's navigation through primary and secondary schooling was treacherous. Despite having received inadequate academic preparation, she enrolled at a community college, a resistive act toward staking a claim of who she was choosing to be in society but stopped out after a semester. Now a mother of three, Kitatoi was determined to act on her environment to achieve her goals of "knowing how the world works" and helping her kids with their schoolwork. Ten years after stopping out, Kitatoi re-enrolled at her local community college as a part-time student. "I felt like that was a big moment for me . . . deciding to go back to school . . . [and] I wanted to tackle the one thing that I was most intimidated by . . . I started with math."

No doubt Kitatoi struggled throughout her academic trajectory, but it should also become evident that she developed a resilient attitude and disposition. Her agentic capability and perception of herself as someone who can succeed was a hard-won standpoint. Her agentic capabilities were rooted in her desire to change her life around—"I want to know how things work . . . know how to write a decent essay. I wanted to be able to know basic math"—and support her children if they one day struggled with schoolwork. Kitatoi's rediscovered agency was once again met with constraining forces, mostly as a result of who gets to participate in the curriculum.

> At the beginning, I felt really embarrassed because I was the oldest one. I was one of the oldest ones there, and I'm learning how to add, subtract, multiply, and divide. . . . I started at basic algebra, basic arithmetic, I learned how to divide at 25. . . . It felt embarrassing . . . even humiliating. But somehow, my mind was just changed like, "I'm going to do this. It's nobody else's life. It's my life." I don't want my kids to come to me and ask for help with long division and I'm not going to be able to help them.

Her action of exercising her agency to take control of events that affect her life led her to tackle her biggest fear. "I decided to start with math because that was my biggest fear in high school. . . . I don't want to repeat the, 'Oh, I'm scared of it' mindset." Restarting her community college journey was an empowering time for Kitatoi; however, it did not come without its challenges. "The first semester that I went, I took a math class, and I really struggled with it. . . . I didn't know how to adjust to doing school and having the kids at the same time." Kitatoi was learning to re-navigate a system that had failed her and left her disenfranchised for ten years. It was evident that there was a whirlwind of environmental influences acting against Kitatoi's agency (i.e., the social stigma of being an older student in college, feelings of embarrassment due to her current circumstance, and a common struggle of navigating the college environment that affects a large portion of first-generation college students).

With the mindset of tackling one of her biggest fears, Kitatoi would enroll in the lowest-level mathematics course available at her community college. In her first year back in college, she would take Pre-Algebra and Elementary Algebra the following semester:

> At first, I wasn't really understanding my classes because . . . I would do all the homework, I would practice the problems, I would stay up late, I would wake up early just to practice more, and I really felt like I was understanding the professor when he taught, but when I took a test, I would get a C. I would get really frustrated with myself like, "I'm really trying my best here," and the best that I could do according to these tests is average.

While agentically acting on her mathematics course, "I was really studying my ass off . . . I'm not studying at an average level"; Kitatoi was again left feeling disempowered, this time by the course content. Her perception of herself as a capable student and a learner was contested. Kitatoi's identity as a competent learner was beginning to take form; during her first semester in college, she was beginning to view mathematics as only for certain types of people.

> I started thinking maybe the saying is true. It's not for everybody. It's only certain type of people that understand this concept. . . . I actually passed that semester of pre-algebra with a C, and I just felt really disappointed . . . because I thought I don't have any more to give. I don't have any more energy to give . . . out of my best effort, all I get is a C . . . I'm not studying at an average level. I'm studying my ass off here, and I'm barely doing average. I started to give up because I felt like, "No, I guess math is just not for me." I passed with a C . . . I was practicing; I was getting help. I was doing all that, and I still wasn't . . . I wasn't getting the As on the exams, because if you get As that means you're really understanding everything.

Kitatoi chose to tackle one of her biggest fears, mathematics, not because in that stage of her life she had a desire to pursue a career in engineering or mathematics but as a result of enacting her agency toward resisting her disenfranchised subject position and taking control of her life's course. While mathematics was "one of [her] biggest fears," Kitatoi made a conscious decision not to allow mathematics course requirements to constrain her academic progression. Based on Kitatoi's prior experience in high school, it seems her desire to get As on exams was a desire to shed the perception of herself as a struggling student or someone who didn't learn anything in high school. Kitatoi's participation in mathematics, in large part, was to reshape how she saw herself, i.e., as a capable learner; unfortunately, her grades were not reflecting who she wanted to be. Nevertheless, Kitatoi demonstrated resilience in progressing through, despite not receiving the performance markers she was expecting.

> And so, I just thought, "Okay, well this is how it's going to be. It's disappointing, but *I guess this is just the type of student than I am*". . . . I felt like a failure. Even though I was back in school doing a good thing . . . I still felt like, "Man, my best isn't good enough."

Although her strategies indicated that she was doing everything to succeed—attending classes, constantly practicing problems, and getting help—there was still a barrier to overcome. When asked about her interactions with her mathematics instructors, Kitatoi stated, "My pre-algebra professor . . . wanted us to learn, would have good explanations and plenty of examples." Learning mathematics and the way students position themselves in relation to mathematics is a trajectory of participation, constructed by both the students and the instructors (Boaler & Greeno, 2000). Kitatoi was in an environment where her willingness to learn was congruent with the instructor's desire for his students to learn. Therefore, although she was not performing at the level she had hoped for, she continued on to the next mathematics course. Boaler and Greeno (2000) posit that the types of mathematical tasks, teaching techniques, and learning approaches used enable or constrain students' beliefs of seeing themselves as math type of people. Students' perceptions of themselves as mathematical learners develop in and through social practice; however, certain environments can have a constraining factor on seeing oneself as a capable mathematics learner, which is where Kitatoi found herself in her second semester in community college:

> My elementary algebra professor was total shit. She yelled; you could tell that she didn't want to be there teaching. She was always grumpy and if you asked questions, she kinda would yell at you . . . she made people feel dumb, we complained about her all the time, and some people reported her to the Dean. Nothing happened, though, that we know of.

When I inquired if this imposed learning environment impacted her, Kitatoi stated that she felt "embarrassed . . . not motivated, hopeless, 'cause I was putting in the work and not getting any positive feedback." Understanding the interaction with Kitatoi's elementary algebra instructor made it clear why she resigned herself to be the type of student who barely gets an average grade in mathematics. She finished her first year in community college in a state of disillusionment by the performance marks she was receiving, lack of positive reinforcement and recognition by her instructor, and belief that *"this is just the type of student than I am"*—thus reinforcing once again the identity of a struggling student. Nevertheless, Kitatoi continued to take mathematic courses, persevering despite an imposed negative learning environment. Her agency to control the events that affect her life, despite environmental constraints, propelled her to enroll in the mathematics course that followed Elementary Algebra. Kitatoi's agency toward resistive acts helped her pushed past the academic push-out and subtractive culture experienced by many Mexican American students (Valencia, 2004; Valenzuela, 1999). Studies have found that a large portion of students who enter community colleges make little progress toward degree completion, specifically noting that completion remains correlated with socioeconomic status (Goldrick-Rab, 2010; Summers, 2003); Kitatoi's agency and intense desire to persevere would prove to be sufficient enough to break down barriers.

> After that year . . . I took an intermediate algebra class, and for some reason when I took that class, it's like . . . I honestly feel like a switch was turned on. . . . I would take an exam, and I got an A. That was my very first A ever. Even in high school, that was my very first A. . . . That was the very first time that I actually felt proud of myself, and I saw that all the studying and the work that I've been putting into the previous math classes, I started seeing it pay off.

Kitatoi's fleeting effort in her first year left her development as a competent mathematics type of person contested; this was evident in her account, "my best isn't good enough," upon receiving Cs even though she "studied [her] ass off." However, by resisting negative beliefs about her capabilities and persevering through, Kitatoi was able to obtain an A in her Intermediate Algebra exam. Thus, her ability to perform her competence in mathematics at the level she was striving for was the turning point, i.e., shifting her self-concept of a struggling mathematics learner to a proud mathematical learner.

> In that class [Intermediate Algebra], I was to the point where I could turn to the student that was next to me or behind me, and I could explain to them what was happening. If they needed help, I was able to help them.

Kitatoi's agency, coupled with her perseverance to learn mathematics, afforded her the ability to perform her competence and eventually come to see herself as a mathematics type of person. While her first year was not reflective of the type of mathematics student she wanted to be, Kitatoi's resilient attitude enabled her to reshape who she was with who she wanted to be: "I had a 'I'm not gonna fail again' attitude . . . I'm gonna get good at it no matter what." Following her experience obtaining "my very first A" and her ability to explain to others what was going on in the class, Kitatoi would go on to take the entire sequence of mathematics courses.

> I moved on to college algebra . . . I took trigonometry. That's when I started working as a math tutor. . . . After trig, I went into calculus, and that's a series of three classes, and I took all of those. I struggled with them also like at the beginning, but . . . I had already learned how to manage my time . . . pretty much how to be a good student so that I can learn.

Kitatoi managed to leave behind the self-concept of a student who struggled and took on the identity of a thriving mathematics learner. Her agency to act against a system that failed her allowed her to reshape her student identity, and she subsequently started to see herself as someone who can do mathematics. Kitatoi eventually constructed an environment where she felt recognized as a competent mathematics learner; she became a tutor for mathematics classes ranging from Pre-Algebra to Calculus III. Kitatoi became involved in tutoring based on her resilient effort to reconstruct her educational trajectory and external recognition.

> I practically lived at the tutoring center, that was my hangout spot and I always went to get help with my homework. . . . One day the guy that ran the tutoring center came up to me . . . he came up to me and was like "hey you should consider tutoring". . . . I was so insecure, shy and in my box that it took that . . . [tutoring center director], to literally make, demand me, in a good way, to do it. I feel like he believed in me when I didn't and I'm very grateful to him.

Being externally recognized as a competent mathematics type of person gave Kitatoi the additive boost to move from a peripheral participant of mathematics to a more fully engaged participant of that community of practice. Now a mathematics tutor, Kitatoi moved past the identity of a struggling student to a competent mathematics type of person who subsequently was receiving the external recognition necessary for identity development (Carlone & Johnson, 2007; Rodriguez, Doran, Sissel, & Estes, 2019). Kitatoi stated, "I was able to explain things . . . and the students would appreciate it. And I was having fun and my confidence was way up . . . it was like a community effort." Kitatoi's entry into the mathematics community of practice allowed her to continue pursuing scientific courses. She would move on to take two chemistry courses and eventually made her way to the calculus-based physics course series. Calculus-based physics courses are considered a gateway into engineering programs (Cass, Hazari, Sadler, & Sonnert, 2011; Tyson, Lee, Borman, & Hanson, 2007; Warne et al., 2019) and support students' beliefs of seeing themselves as engineers (Godwin et al., 2016; Verdín, Godwin, Sonnert, & Sadler, 2018). To solidify her transition into an engineering bachelor's degree granting program, Kitatoi needed to pass through this course series.

> I thought, "You know what, I think mechanical engineering would be something that I'm interested in since I'm really curious about how things work and how things are made". . . . I knew I had to take a phys- ics class . . . but I was intimidated by physics. . . . I was just still doubt- ing myself . . . I . . . just left physics to the end because I honestly just thought I was going to fail at it, because only smart people did physics.

Studies have documented that preparation in mathematics is a strong pre- dictor of success in physics (Hazari, Tai, & Sadler, 2007; Kost, Pollock, & Finkelstein, 2009), and while Kitatoi had established herself as a competent mathematics learner, her ability to begin engaging in physics courses was met with environmental constraints. Her statement, "only smart people did physics", was in response to the pervasive gender imbalance in the physics courses, the constant environmental message of who can participate, the stereotype threat that continues to impinge on women's performance, and implicit bias against women's ability to excel in physics courses (Blackburn, 2017; Eddy & Brownell, 2016). Kitatoi further clarified her assumption of who the "smart" people were.

> I worked at the tutoring center . . . at that time. . . . I would see the people that would go into the tutoring center for . . . physics and they were just really smart people . . . meaning they get straight As . . . really smart and mostly guys . . . these younger people that are taking these physics classes, you just need so much dedication, and you just need to study all the time and that's pretty much your life, and I can't do that. Because again, I have three kids.

Compared to other science courses, e.g., biology or chemistry, the gen- der disparities between students who participate in calculus-based physics courses are significantly pronounced. Additionally, studies have found that women often feel a lack of belonging in these courses compared to men, in large part as a result of the gender imbalance and male-dominated culture (Banchefsky, Lewis, & Ito, 2019; Lewis, Stout, Pollock, Finkelstein, & Ito, 2016). In general, women are less likely to participate in the calculus-based physics series, and this persistent culture of exclusion among certain disci- plines makes for the heightened invisibility of women. Kitatoi's confidence in mathematics was only a starting point for her trajectory into engineering;

seeing herself as a physics type of person was equally important for her internal recognition as a capable engineering type of person.

> Then I went into physics, and it was just like . . . out of everything that I have taken in college . . . [physics] was way more fascinating. . . . I just got really fascinated with a lot of things that I didn't know. I didn't know pretty much how the world goes around, how things move, how things interact with each other, what things are made out of, how does the car engine work, how does gravity work, stuff like that. . . . [W]hen you get to those classes, you see what's really happening . . . I just started getting really fascinated with all those things. Then I decided I was going to try mechanical engineering because the most that I was interested in is how things worked. How are things made? What are they made out of? How do they work? If they break, how do they get fixed?

Kitatoi's six-year journey through community college was permeated with up and down moments; through her resilient agentic capabilities, she persevered to degree completion, earning three associate's degrees in multiple sciences, mathematics, and physics. While it is evident that Kitatoi's trajectory into engineering was anything but linear, taking on the identity of a mathematics and physics type of person was instrumental for her pathway into engineering. Kitatoi's ability to break away from the academic self-concept of a struggling student to someone who was competent in mathematics was instrumental in enabling her agency to pursue other STEM courses. Prior studies have found that students' exposure and affinity toward mathematics and physics are gateways into seeing oneself as the type of person who can do engineering (Cribbs, Cass, Hazari, Sadler, & Sonnert, 2016; Godwin et al., 2016; Verdín, Godwin, & Ross, 2018; Verdín, Godwin, Sonnert, et al., 2018).

Kitatoi is now enrolled at a four-year university in the mechanical engineering program. I end this chapter with an acknowledgment that Kitatoi's six-year journey through community college, although arduous at times, was an empowering experience where she managed to shed the academic self-concept of a struggling student and learned to see herself as part of the math and physics community of practice. Entry into these communities of practice was significant for Kitatoi's belief of seeing herself as someone who can do engineering.

## Discussion

In Kitatoi's narrative, the imposed environmental constraints in the education system when she was an adolescent had a long-lasting impact on how she saw herself. Kitatoi's resilient agentic capabilities propelled her to change the trajectory of her life's course. However, the route to engineering is layered with hurdles; students must simultaneously see themselves as mathematics and physics type of people to feel as though they can do engineering (Godwin et al., 2016; Verdín, Godwin, & Ross, 2018). By enacting her agency and resiliently pushing past the environmental constraints that had left her as an academically ill-prepared adult, Kitatoi developed an identity as a math and physics person, and these identities supported her belief of being capable of pursuing an engineering career path. However, her journey through community college was challenged by a negative teacher–student interaction and by implicit gender bias of who gets to participate in the STEM curriculum. Kitatoi's identity of being a woman and a mother seemed at odds with the male-dominated makeup of the calculus-based physics courses and general culture of these courses. She came to see herself as someone who can do engineering through the performance and competence beliefs she developed in mathematics, through the recognition she received by the mathematics tutoring director that subsequently led her to become a math tutor. Kitatoi's decision to enroll in the calculus-based physics courses was an agentic resilient act intended to respond to the imposed male-dominated environment and a response to her interest in learning how the world works. The identity-forming experiences throughout Kitatoi's community college trajectory, in the context of mathematics and physics, supported her decision to pursue a career in engineering.

Importantly, Kitatoi's counter-narrative emphasizes how an engineering career pathway is still open to nontraditional students who are relearning basic arithmetic skills. All students entering community college, despite their current mathematical skills, should be encouraged to explore the options of a career in engineering. Students may be entering community colleges with preconceived notions of their academic capabilities, based on challenges or demoralizing experiences from their prior schooling. However, community college educators and counselors have the opportunity to create a space where adult learners can begin to foster disciplinary role identities (i.e., in the mathematics and physics context). The brief by Rodriguez et al. (2019) offers a breadth of examples of current programmatic strategies that can support community college identity development. However, for those who are full-time parents, students, and also have a job, programmatic opportunities can add more work on top of their busy schedules. Recognition as capable STEM learners by educators, counselors, and staff is important for students' motivation and identity development, but it is not the only factor. Educators could also help students shift their attitudes and beliefs about their STEM capabilities by sharing their struggles and the resiliency of their effort toward achieving their goals. Lastly, understanding community college students' trajectories through an agency lens shifts the victim-blaming discourse and deficit outlook of these groups of students.

## Note

1. Pseudonym was selected by the participant due to sentimental value.

# References

Ballón, E. G. (2008). Racial differences in high school math track assignment. *Journal of Latinos and Education, 7*(4), 272–287. doi:10.1080/15348430802143428

Ballón, E. G. (2015). *Mexican Americans and education: El saber es poder.* Tucson, AZ: University of Arizona Press.

Banchefsky, S., Lewis, K. L., & Ito, T. A. (2019, October). The role of social and ability belonging in men's and women's pSTEM persistence. *Frontiers in Psychology, 10,* 1–16. doi:10.3389/fpsyg.2019.02386

Bandura, A. (2001). Social cognitive theory: An agentic perspective. *Annual Review of Psychology, 52*(1), 1–26. doi:10.1146/annurev.psych.52.1.1

Blackburn, H. (2017). The status of women in STEM in higher education: A review of the literature: 2007–2017. *Science and Technology Libraries, 36*(3), 235–273. doi:10.108 0/0194262X.2017.1371658

Boaler, J., & Greeno, J. G. (2000). Identity, agency, and knowing in mathematics worlds. In J. Boaler (Ed.), *Multiple perspectives on mathematics teaching & learning* (pp. 171–200). Westport, CT: Greenwood. doi:10.1109/CEIDP.1997.641133

Carlone, H. B., & Johnson, A. (2007). Understanding the science experiences of successful women of color: Science identity as an analytic lens. *Journal of Research in Science Teaching, 44*(8), 1187–1218. doi:10.1002/tea

Cass, C. A. P., Hazari, Z., Sadler, P. M., & Sonnert, G. (2011, June 26–29). *Engineering persisters and non-persisters: Understanding inflow and outflow trends between middle school and college.* Paper presented at the American Society for Engineering Education (ASEE) Annual International Conference, Vancouver, BC, Canada.

Cribbs, J. D., Cass, C., Hazari, Z., Sadler, P. M., & Sonnert, G. (2016). Mathematics identity and student persistence in engineering. *International Journal of Engineering Education, 32*(1), 163–171.

Cruz, J., & Kellam, N. (2018). Beginning an engineer's journey: A narrative examination of how, when, and why students choose the engineering major. *Journal of Engineering Education, 107*(4), 556–582. doi:10.1002/jee.20234

Eddy, S. L., & Brownell, S. E. (2016). Beneath the numbers: A review of gender disparities in undergraduate education across science, technology, engineering, and math disciplines. *Physical Review Physics Education Research, 12*(2), 1–20. doi:10.1103/PhysRevPhysEducRes.12.020106

García, E. E., Wiese, A.-M., & Cuéllar, D. (2011). Language, public policy, and schooling: A focus on Chicano English language learners. In R. R. Valencia (Ed.), *Chicano school failure and success: Past, present, and future* (3rd ed., pp. 143–159). New York, NY: Routledge.

Gee, J. P. (2001). Identity as an analytic lens for research in education. In W. G. Secada (Ed.), *Review of research in education* (Vol. 25, pp. 99–126). Washington, DC: American Educational Research Association.

Godwin, A., Potvin, G., Hazari, Z., & Lock, R. (2016). Identity, critical agency, and engineering: An affective model for predicting engineering as a career choice. *Journal of Engineering Education, 105*(2), 312–340. doi:10.1002/jee.20118

Goldrick-Rab, S. (2010). Challenges and opportunities for improving community college student success. *Review of Educational Research, 80*(3), 437–469. doi:10.3102/0034654310370163

Gonzalez, G. G. (2013). *Chicano education in the era of segregation.* Denton, TX: University of North Texas Press.

Hazari, Z., Sonnert, G., Sadler, P. M., & Shanahan, M.-C. (2010). Connecting high school physics experiences, outcome expectations, physics identity, and physics career choice: A gender study. *Journal of Research in Science Teaching, 47*(8), 978–1003. doi:10.1002/tea.20363

Hazari, Z., Tai, R. H., & Sadler, P. M. (2007). Gender differences in introductory university physics performance: The influence of high school physics preparation and affective factors. *Science Education, 88*(1), 25–32. doi:10.1002/sce

Holland, D., Lachicotte, W., Skinner, D., & Cain, C. (1998). *Identity and agency in cultural worlds.* Cambridge, MA: Harvard University Press.

Johnson, A., Brown, J., Carlone, H., & Cuevas, A. K. (2011). Authoring identity amidst the treacherous terrain of science: A multiracial feminist examination of the journeys of three women of color in science. *Journal of Research in Science Teaching, 48*(4), 339–366. doi:10.1002/tea.20411

Kost, L. E., Pollock, S. J., & Finkelstein, N. D. (2009). Characterizing the gender gap in introductory physics. *Physical Review Special Topics-Physics Education Research, 5*(1), 1–14. doi:10.1103/physrevstper.5.010101

Lewis, K. L., Stout, J. G., Pollock, S. J., Finkelstein, N. D., & Ito, T. A. (2016). Fitting in or opting out: A review of key social-psychological factors influencing a sense of belonging for women in physics. *Physical Review Physics Education Research, 12*(2), 1–10. doi:10.1103/PhysRevPhysEducRes.12.020110

Oyserman, D., Elmore, K., & Smith, G. (2012). Self, self-concept, and identity. In M. R. Leary & J. P. Tangney (Eds.), *Handbook of self and identity* (2nd ed., pp. 69–104). New York, NY: Guilford Press.

Pruyn, M. (1999). *Discourse wars in Gotham-West: A Latino immigrant urban tale of resistance & agency.* New York, NY: Routledge. doi:10.4324/9780429048357

Rodriguez, S. L., Doran, E. E., Sissel, M., & Estes, N. (2019). Becoming la ingeniera: Examining the engineering identity development of undergraduate Latina students. *Journal of Latinos and Education,* 1–20. doi:10.1080/15348431.2019.1648269

Rodriguez, S. L., Hensen, K. A., & Espino, M. L. (2019). Promoting STEM identity development in community colleges & across the transfer process. *Journal of Applied Research in the Community College, 26*(2), 11–21.

Rossiter, M. (1999). A narrative approach to development: Implications for adult education. *Adult Education Quarterly, 50*(2), 56–71. doi:10.1146/annurev.genom.2.1.343

Rumberger, R. W., & Rodríguez, G. M. (2010). Chicano dropouts. In R. R. Valencia (Ed.), *Chicano school failure and success: Past, present, and future* (3rd ed., pp. 76–98). New York, NY: Routledge.

Sapp, V. T., Kiyama, J. M., & Dache-Gerbino, A. (2015). Against all odds: Latinas activate agency to secure access to college. *NASPA Journal About Women in Higher Education, 9*(1), 39–55. doi:10.1080/19407882.2015.1111243

Schwartz, S. J., Luyckx, K., & Vignoles, V. L. (2011). *Handbook of identity theory and research.* New York, NY: Springer.

Solorzano, D. G., & Solorzano, R. W. (1995). The Chicano educational experience: A framework for effective schools in Chicano communities. *Educational Policy, 9*(3), 293–314. doi:10.1177/0895904895009003005

Solorzano, D. G., & Yosso, T. (2000). Toward a critical race theory of Chicana and Chicano education. In C. Tejeda, C. Martinez, & Z. Leonardo (Eds.), *Charting new terrains of Chicana (o)/Latina (o) education* (pp. 35–65). New York, NY: Hampton Press.

Solórzano, D. G., & Yosso, T. J. (2002). Qualitative inquiry framework for education research. *Qualitative Inquiry, 8*(1), 23–44. doi:10.1177/107780040200800103

Stanton-Salazar, R. D. (2001). *Manufacturing hope and despair: The school and kin support networks of US-Mexican youth.* New York, NY: Teachers College Press.

Summers, M. D. (2003). ERIC review: Attrition research at community colleges. *Community College Review, 30*(4), 64–84. doi:10.1177/009155210303000404

Tejeda, C., Martinez, C., & Leonardo, Z. (Eds.). (2000). *Charting new terrains of Chicana(o)/Latina(o) education.* New York, NY: Hampton Press.

Tyson, W., Lee, R., Borman, K. M., & Hanson, M. A. (2007). Science, technology, engineering, and mathematics (STEM) pathways: High school science and math coursework and postsecondary degree attainment. *Journal of Education for Students Placed at Risk (JESPAR), 12*(3), 243–270. doi:10.1080/10824660701601266

Valencia, R. R. (Ed.). (1997). *The evolution of deficit thinking: Educational thought and practice.* Washington, DC: Falmer Press.

Valencia, R. R. (2004). *Chicano school failure and success: Past, present, and future.* New York, NY: Routledge.

Valencia, R. R. (2011). The plight of Chicano students: An overview of schooling conditions and outcomes. In R. R. Valencia (Ed.), *Chicano school failure and success: Past, present, and future* (3rd ed., pp. 3–41). New York, NY: Routledge.

Valencia, R. R., & Black, M. S. (2019). "Mexican Americans don't value education!" On the basis of the myth, mythmaking, and debunking. In E. G. J. Murillo (Ed.), *Critical readings on Latinos and education* (pp. 3–24). New York, NY: Taylor & Francis.

Valenzuela, A. (1999). *Subtractive schooling: U.S.-Mexican youth and the politics of caring.* Albany, NY: State University of New York Press.

Verdín, D., Godwin, A., Kirn, A., Benson, L., & Potvin, G. (2019). Engineering role identity fosters grit differently for women first- and continuing-generation college students. *International Journal of Engineering Education, 35*(4), 1037–1051.

Verdín, D., Godwin, A., & Ross, M. (2018). STEM roles: How students' ontological perspectives facilitate STEM identities. *Journal of Pre-College Engineering Education Research (J-PEER), 8*(2), 31–48. doi:10.7771/2157-9288.1167

Verdín, D., Godwin, A., Sonnert, G., & Sadler, P. M. (2018, October 3–6). *Understanding how first-generation college students' out-of-school experiences, physics and STEM identities relate to engineering possible selves and certainty of career path.* Paper presented at the IEEE Frontiers in Education (FIE) Conference, San Jose, CA.

Warne, R. T., Sonnert, G., & Sadler, P. M. (2019). The relationship between advanced placement mathematics courses and students' STEM career interest. *Educational Researcher, 48*(2), 101–111. doi:10.3102/0013189x19825811

Yosso, T. J. (2013). *Critical race counterstories along the Chicana/Chicano educational pipeline.* New York, NY: Routledge.

# 10 Leadership Through the Lenses of Latinas

## Undergraduate College Students in STEM-Related Disciplines at Regional HSIs

*Hilda Cecilia Contreras Aguirre, Rosa Banda, and Elsa M. Gonzalez*

Latinas' enrollment in postsecondary education has improved in disciplines such as social sciences and psychology (Mitts, 2016; Schoon, 2015); however, Latinas' degree attainment in computer science, mathematics, physics, and engineering has remained low (National Science Foundation [NSF], 2016). For example, in 2014, only 2.2% of Latinas obtained a college degree in science, technology, engineering, and mathematics (STEM) fields (NSF, 2017). In particular, Latina college students at an undergraduate level often face adversity, which has included but was not limited to low socioeconomic and first-generation status, racism, and feelings of isolation (Cabrera & Padilla, 2004). Moreover, Onorato and Musoba (2015) argue that even Hispanic-serving Institutions (HSIs), whose purpose is to better serve minoritized students, still operate with a White frame approach and influence minoritized female students' leadership. In this regard, Haber-Curran, Miguel, Shankman, and Allen (2018) found that women have the ability to build leadership capacity by developing relationships. These connections provide value and strength to leadership development which, in turn, female college students can use to thrive in STEM.

This qualitative study was conducted in the summer and fall of 2018 with the participation of ten Latina undergraduate students. The principal researcher collected and triangulated data via interviews, observations, and analysis of documents in a six-month timeline. Participants' narratives enriched our understanding of the strategies, activities, and perceptions that influenced how Latinas adopted leadership styles as undergraduate STEM college students. Often, Latinas' leadership approach as women and ethnic minoritized students can be misunderstood or disregarded; however, Latinas provided numerous examples of initiatives and attempts to create better college environments for other Latina and younger students. On multiple occasions, participants noted the criticality to assume leadership roles within campus organizations and programs. Such experiences, the participants mentioned, were informative to acquire the skills and occupy leadership positions to later assist Latinas in their profession. This is just one of the multiple strategies that Latinas can use as a form of resilience to be successful STEM students. The development of resiliency is dependent on many key factors, which include support (extrinsic) and individual (intrinsic) motivation (Greene, 2002). Other resilience factors that contribute to the individual include potency, stamina, and personal causation (Van Breda, 2001).

This chapter begins with a brief overview of leadership as it relates to ethnicity and gender. It is relevant to comprehend the interconnection among all three aspects (ethnicity, gender, and leadership) and understand Latinas' unique approach to leadership. Then, we present a theoretical framework based on the leadership labyrinth model which focuses on human capital, gender differences, and discrimination toward women. Next, we introduce the methods including three subsections—data collection, data analysis, and trustworthiness. Subsequently, we present the study's findings. Finally, a discussion section yields implications for staff and faculty to support and reinforce Latinas' leadership development. In the end, we note a few recommendations for further studies.

The purpose of this study was to gain insight into the leadership practices adopted by Latina college students when they pursue STEM disciplines in mostly male-dominated majors at two regional HSIs. The primary research question is:

1. What kind of activities, if any, do Latina college students perform that show leadership styles in STEM-related disciplines?

The ancillary research questions are:

1. What are the leadership strategies, if any, that Latina college students use to successfully navigate their college experience in male-related disciplines?
2. What type of experiences, if any, influence the leadership adopted by Latina college students who pursue degrees in male-dominated fields?

## Literature Review

### *Leadership and Latina Ethnicity*

Latinos/as' leadership perceptions may differ from the mainstream conceptualization of leadership. As such, a leader can be a person who is capable and has the willingness to be hardworking regardless of their hierarchical status (Lozano, 2015). In college, Beatty (2015) noted the importance of student organizations to help Latino/as create bonds with other students, develop a sense of belonging, and enhance their participation in campus-wide events. Lozano (2015) further emphasized the importance of aspects such as "collaboration, community, and empowerment" (p. 11) to shape and define a Latino/a leadership identity. In turn, Suarez (2015) reflected on the lack of participation of Latinos/as in leadership workshops and programs that is mainly due to the irrelevant connection between Latinos/as and faculty (including staff). Accordingly, this lack of connection does not allow a genuine and meaningful interaction between students and university personnel. In addition, Latinos/as feel that such "leadership opportunities did not include a Latino culture perspective" (Suarez, 2015, p. 35). Therefore, Latinos/as' lack of participation in these events showed an alarming lack of connection to Latino culture.

Despite facing difficulties and hostile college climates, Latinas learn to lead with a collaborative effort underlying the importance of communal work, Latinos/as' toughness, and cultural heritage (Foulis, 2017). Through involvement in student organizations, Latinas develop synergies to counteract racism while strengthening "social activism" (Beatty, 2015, p. 51). As such, the aspect of activism is present when Latinas lead as they try to empower other women and Latinas (Onorato & Musoba, 2015), highlighting the need for "a community-based model of leadership" (Lozano, 2015, p. 12). According to Guardia (2015), a community-oriented group is found through Latina sororities that promote and develop an ethnic identity, highlighting the aspects of language and community. More importantly, such organizations provide opportunities for on-campus involvement and leadership development (Guardia, 2015).

### Leadership and Gender

Historically, the inequalities between dominants and subordinates have shaped male and female relationships (Miller, 1986). Kezar and Moriarty (2000) noted that women's and minorities' perceptions of leadership differ from those of men and ethnically dominant elites. According to Miller (1986), such male superiority reflects a perpetual inequality whose aim is to maintain males' power and dominance. Women usually are highly involved in developing cooperation, promoting others' creativity, and helping others to grow; such activities are important but unvalued in a male-led social structure (Miller, 1986). As such, women and minorities participate in social-related activities as members of student organizations and develop leadership skills from a collaborative and interrelated perspective (Kezar & Moriarty, 2000).

Women recognize the need for cooperation to "aid and enhance the development of other human beings while advancing one's own" (Miller, 1986, p. 41). In this regard, Jordan, Hartling, and Walker (2004) pointed out that usually, women care about both their own needs *and* the needs of others. As a result, women highly appreciate establishing affiliations with other individuals, including helping them to grow while neglecting even their own enhancement (Miller, 1986). In particular, women pay special attention to the actions and feelings they generate in others as a consequence of their interactions. Because women possess "relational awareness" (Jordan et al., 2004, p. 14), they are able to acknowledge simultaneously what happens with themselves, others, and their relationships. Contrarily, men generally focus on their own needs using a "single voice discourse" (Jordan et al., 2004, p. 14).

Indeed, leadership "is not gender, race, and ethnicity neutral" (Onorato & Musoba, 2015, p. 30). Therefore, individuals' gender and culture affect the type of leadership adopted that ultimately develops a leadership identity. As such, the increasing number of Latinas in higher education calls for a better understanding of minoritized women's leadership experiences in college. Additionally, the low representation of Latina students in some STEM disciplines, even at HSIs, requires the examination of appropriate resources and programs that promote leadership development. Lastly, the acquisition of leadership skills has been proven to be fundamental as core knowledge in STEM to accomplish academic goals and later, organizational objectives (Kendricks, Arment, Nedunuri, & Lowell, 2019). In addition, Haber-Curran et al. (2018) noted that when women build purposeful relationships with a leadership mindset, they can be aware of the value that leadership can have in their lives.

## Theoretical Framework

The leadership labyrinth model (Eagly & Carli, 2007) undergirds this study. The labyrinth resembles the difficulties and problems that women face to become leaders (Eagly & Carli, 2007). A leader, according to Eagly and Carli (2007), is a "person who exercises authority over other people" (p. 8), who can influence and motivate others, and who helps with the organization and coordination of a given task. The occupation of leadership positions has not reached women and men equally, not even in modern nations (Kark & Eagly, 2010). In addition, some ethnicities such as Whites remain overrepresented in leadership positions across the globe (Feagin, 2013).

In this regard, Eagly and Chin (2010) argue the importance of the intersectionality of race and ethnicity when it comes to adopting a leadership style, which often triggers discrimination against minoritized populations. Therefore, individuals' leadership is not only strongly shaped by their ethnicity but also influenced by their gender. Kark and Eagly (2010) went further by pointing out how minorities have the advantage of considering different perspectives when it comes to leading, including those of the majority. In particular, flexibility and the ability to negotiate are two important skills that women as minorities learn in their journey to become leaders (Kark & Eagly, 2010). In addition to gender, Latinas encounter social pressure to adopt a leadership approach aligned with the dominant group leadership traits (Eagly & Chin, 2010), especially in contexts where Latinas are usually not considered leaders (Eagly, 2007).

Furthermore, Latinas' leadership is important when it comes to using cultural concepts and knowledge along with strong advocacy for others (Eagly & Chin, 2010). Therefore, the leadership labyrinth model helps explain Latina college students' difficulties and abilities to lead—for instance, highlighting the special skills of women's leadership, whose ability to lead includes a more collaborative approach (Eagly, 2007). In some cases, women's commitment and actions go beyond their leadership functions, as they often develop a friendship with followers or become their mentors. The aforementioned example makes clear the difference between women's and men's leadership approaches (Eagly, 2007).

## Methods

The method of this qualitative research consists of using a case study approach (Yin, 2009) at two Hispanic-serving Institutions (HSIs). The study included ten Latina undergraduate students pursuing STEM majors to examine the type of activities, if any, Latinas perform that show leadership styles in STEM male-dominated disciplines. A purposeful sample was utilized, and the selection criteria to participate in the study were as follows: (1) Latina undergraduate student; (2) enrolled in their senior year; (3) pursuing chemistry, computer science, electrical engineering, geology, mathematics, mechanical engineering, or physics; and (4) attending one of the two selected HSIs. These two HSIs were selected due to similar patterns in enrolling a large number of Latinas, but they both had a critical underrepresentation of Latinas in the aforementioned majors. Data triangulation consisted of interviews with Latina college students, demographics sheets, observations from academic settings, and analysis of documents of Latinas' academic performance at these two research sites (Erlandson, Harris, Skipper, & Allen, 1993; Lincoln & Guba, 1985).

### Data Collection

To collect data, a semi-structured interview protocol was employed with participants. Interviews were conducted at students' respective university libraries throughout the summer and fall of 2018. In addition, the primary researcher maintained a reflexive journal during the time of data collection and data analysis to document observations, thoughts, logistic changes, and unexpected setbacks (Glesne, 2011). Lastly, the document analysis, which included institutional statistics of Latinas' persistence, helped in understanding Latinas' intentions to major in STEM, persistence in the STEM programs, and close degree attainment. Participants' identification was confidential; hence, students self-selected pseudonyms for the study.

*Table 10.1* Themes and Categories of the Study.

| Themes | Coming Out of the Shadows | Embracing Their Identity as Women and Latinas |
|---|---|---|
| **Categories** | Gaining visibility<br>Building a social network | Fighting to be heard<br>Feeling proud of who they are |

## Data Analysis

Data were examined through inductive analysis (Lincoln & Guba, 1985). Interview data were transcribed verbatim, yielding the identification of units of data with Dedoose, an online platform to help in qualitative data analysis. Next, the process of codification encompassed sorting data units multiple times until a framework of themes and categories emerged (Glesne, 2011). The principal researcher then categorized information from all ten interviews under categories, which attached to themes. Table 10.1 shows the themes and categories that emerged from data analysis.

## Trustworthiness

In qualitative inquiry, researchers should address aspects such as confidence in the findings, applicability in other settings, consistency to repeat the study, and impartiality of findings (Lincoln & Guba, 1985) to assure the trustworthiness in the study. Therefore, the following techniques guarantee the study's reliability. To ensure *credibility*, member checks included participants' involvement, as each of them had an opportunity to review and edit her interview transcript. In addition, observations conducted at research site facilities and document analysis helped in the triangulation of data. To ensure *transferability*, demographic sheets provided thick description (Lincoln & Guba, 1985) of participants, and purposeful sampling helped determine the selection criteria, which could lead researchers to recruit a similar group of participants. To ensure *dependability and confirmability*, a reflexive journal employed throughout the study was utilized as a self-reflective instrument and a tool to make decisions regarding the study.

## Findings

Participants proudly shared the multiple experiences in which they were willing to support other students, who sometimes were younger peers, women, and Latinas. In this sense, most participants were fully involved in assuming leadership roles in on-campus student organizations. Other activities include accompanying and advising freshman students in college/major-related topics through mentoring positions, reaching younger college students to communicate research opportunities, and participating in outreach activities to persuade young female students in high school to pursue STEM majors. The aforementioned activities proved participants' involvement and commitment within the campus community, which open spaces for Latinas to exercise leadership. The following themes emerged: (1) Latinas' strategies to overcome difficulties in college through *Coming Out of the Shadows*; and (2) Latinas' discernment to skillfully assume attitudes related to leadership by means of *Embracing Their Identity as Women and Latinas.*

### Coming Out of the Shadows

Participants' presence in male-dominated classrooms comes with some difficulties. Mainly, these troubles have to do with Latinas' interactions with their male peers, who undervalue the intellectual capacity in STEM. Such a struggle makes Latinas espouse alternative approaches. For example, in classrooms, Latinas participate more during lectures, develop alliances with other women or Latinas, guide in-class projects, and voice their opinions, especially if they perceive unfairness. Outside of the classroom, Latinas occupy important positions in student organizations, develop community-based projects, perform jobs in which they inspire and motivate other peers to do well in college, and contribute to generating a learning community on campus. As such, the categories framing this theme are as follows: *Gaining visibility* and *Building a social network.*

*Gaining visibility.* Through participants' college pathways, they were aware of the importance of their involvement in programs and organizations. Such participation allowed them to interact and communicate with others, including faculty, staff, and students. The experiences of participants as it relates to assuming leadership positions in students' organizations, campus events, and in-class activities, to mention a few, contributed to enhancing their visibility and self-reliance. Celina mentioned having diverse roles in such organizations:

> I'm a part of the National Society of L and S. I am a Success Network-ing facilitator, so I mentor students, who are joining the society, set goals and work towards them throughout the semester. I was recently elected [name of organization] Junior chair, so I am part of the Society H Engineers and what I will be doing is going to high schools and encouraging students to pursue STEM degrees and helping their parents start applying for college, FAFSA and scholarships.

Similar to Celina's experience participating in events motivating students to pursue STEM majors, Isela's job on campus as a mentor is encouraging new students to continue with their majors. She stated, "I work with the computer science students, so I can help them specifically with my experiences that I went through my first year in computer science." Both Celina and Isela recognize the need to integrate other students to STEM by showing their experience in their field. In turn, Isela showed her commitment to initiating other women in STEM. In this regard, she expressed:

> I actually worked with her [a professor] over the summer, for girls coding camp which was really fun because I got to encourage some, like, little girls to try STEM and hopefully they like it in future, and they might come study STEM, like computer science. I really hope they do.

By experiencing the lack of women in STEM-related fields, Isela happily revealed her satisfaction in working with younger girls to inspire and to motivate them to pursue STEM degrees. Similarly, according to Sina, it is very important to interact with new students, especially freshman and sophomore engineering students. Because new students need role models, senior students can take that role to help newer students develop a sense of belonging in the college learning community. Sina used the words "inspiration" and "motivation" to help students move forward in difficult disciplines. Karina showed her leadership with her campus' involvement. Karina said:

> I'm very involved in various organizations on campus. I am the president for [name of organization] . . . it's predominantly male, when I became president there were only two females and now we have seven, but it's still predominantly male, and I'm also the vice president for [name of organization], and I'm very involved with the engineering community and helping out others.

Karina expressed her interest in helping other students on campus, and she discovered that taking a leadership role in multiple organizations could be a way to highlight women's presence. All the aforementioned participants put into practice a set of abilities they considered essential to influence others—either younger students or other females—in pursuing STEM disciplines.

*Building a social network.* To be leaders, participants needed to interact with other people; these people were often their peers with whom they developed a special connection. Having such a social network allows participants to operate in challenging classrooms, with difficult class subjects, and their particular ability to thrive. Sina defined her peer support system; she said, "Your friends come in your classes to create that support system here because engineering is really hard on its own and being able to manage it and not lose your mind is really difficult." Sina's perspective provided an insight into how valuable college peers are for Latinas pursuing STEM disciplines, to both make it easy and surpass challenges. The constant exchange among peers reveals a collective effort to succeed academically; through friendship with peers, Latinas find ways to thrive and excel in college. In the same sense, Karina noted how important her classmates were for her to foster a campus learning community, saying, "I like the small campus, I like knowing all my classmates, we're pretty much together for all the years that we are here. It is a good community." Karina's statements made clear that small-sized campuses have the advantage of allowing students to build peer support networks that contribute to students' sense of belonging and academic commitment.

Similarly, Emily verbalized her experience with her classmates with whom she has developed strong bonds due to personal and academic similarities. She noted, "We have been in like the same classes since freshman year, so grew together doing this, working toward the same goal is motivating, you're not alone either, you don't feel so alone in this, in gaining my degree." In the aforementioned comment, Emily reflected on the importance of her classmates to succeed in college, especially because she relies a lot upon them for studying and doing homework. In turn, Sina commented, "In our class . . . we always help each other out on labs or homework. . . . We collaborate to find solutions together which engineering is all about." Sina, Celina, and Emily mentioned that collaborative work helps them face challenges, find strategies, and encourage each other. These testimonies show the importance of communal help for Latinas in STEM fields.

### Embracing Their Identity as Women and Latinas

According to participants, being a STEM college student, especially in disciplines where females and Latinas are underrepresented, is not easy. Participants realized that being the only female in classrooms, or being one of just a few females, was challenging in terms of how White male peers perceived them. In addition, participants' heritage as it relates to their culture, values, and ethnicity helped them operate in demanding college environments. All the aforementioned factors contributed to developing participants' leadership attitudes and reinforcing how other people in and out of college perceive them. Two categories frame this theme: *Fighting to be heard* and *Feeling proud of who they are.*

*Fighting to be heard.* Participants were underrepresented in STEM disciplines both in terms of their gender and ethnicity. They found it difficult to interact with their male peers. Such interactions were often conflictive and disrespectful, sometimes through implicit actions and often with explicit behaviors. For the most part, participants found inner strength as women in dealing with academic and gender-related issues. Karina stated, "I am president and vice president [of student organizations' names] to show [male peers] that we are equal if not better at some things." Furthermore, participants' leadership was found in classrooms, where they have to be able to face some negative male mindsets. Sina insisted:

They [male classmates] come into this thinking that they are going to walk all over us, and, luckily, the females that we have in our classes we push back, and we take leadership positions in our classes. I mean, I'm a secretary for [name of organization]; M. is the president. All of our female students are involved in our classes, we stand as a team and make sure we do not get pushed around.

Sina's statement shows how women can be leaders in engineering classrooms. This suggests that Sina's (Latina) peers do not let themselves be intimidated by their male colleagues in classrooms. More importantly, they actively voice their opinions and take key leadership roles to feel more represented in engineering. Similar to Sina, Karina observed, "A lot of women in our college . . . are stepping up because they all feel like they have a lot to prove in this male-dominated world." Both Sina and Karina notice the need to become more visible in classes, be more vocal, and challenge their classmates to perform better. In addition, Alyssa's participation in an honors program helped her learn about internships, scholarships, and even job opportunities. Such opportunities are perceived by Alyssa as an advantage to stand out as a student. Moreover, Alyssa pointed out her interest in being involved in student life in college:

I really like to be involved on campus, so I am a college ambassador, which is being a campus tour guide, honors program. I'm in student volunteer connection, you pretty much do volunteer work and service events throughout the year. I do a lot of holidays, A. Days every summer, which is a transitional camp for incoming students, I recently joined the N. Society of leadership and . . . and then also, I recently joined a Hispanic Engineers organization.

Alyssa also participated in an (ethnic and field-related) conference, at which female engineering students talked about gender-related difficulties in male-dominated fields. Overall, participants found different strategies to cope with negative male classmates' interactions. They believe firmly in their self-efficacy and resiliency to be Latinas in STEM.

*Feeling proud of who they are.* Throughout the interviews, participants reflected on their peculiarity as women and Latinas. They highlighted the importance of bringing back aspects, such as being aware of their previous experiences in education, their cultural legacy, and their inner strength as women. For Isela, joining a cultural organization allowed her to know more about her Hispanic heritage and other cultural-related topics. Like Isela, Brianna's perspective on Hispanic heritage on campus was influenced by her participation as the director of a cultural association. Brianna noted, "It's a lot more than just Hispanic. We do like different cultures, so I really enjoy that because I love seeing different parts of the world and how people are so different, but yet so much the same." This suggests that despite differences, cultures are rooted in similarities in spite of nuanced experiences; the latter is what remains distinctive in nature. Both Isela and Brianna, with their participation and leadership in this type of cultural organization, feel more culturally aware and able to better understand others who are different from them.

Being a minority in STEM disciplines, specifically in engineering, comes with the challenge for women to find a place where they can also lead, propose, and innovate. According to Sina, "Having that idea of you (women) should be in the shadow, it doesn't work. In engineering [it] is very difficult to get through if you aren't willing to develop the skills that you need to be a leader." According to Sina, both males and females have the same ability to become leaders. In this same subject, Arianne noticed, "They (male peers) would think that you don't know stuff . . . I felt it sometimes . . . I can be as smart as you are, and you will be surprised." Arianna seems to struggle to understand her male peers' behavior in classrooms; however, she firmly believes in her inner strengths and motivation. In turn, Karina is trying to reach other Latinas and trigger their inspiration to consider a career in STEM. She argued:

> I am trying to set a platform that I want to be a role model, I want to show other Latinas, also make sure anytime we have like an outreach to go to high schools . . . I want to make sure I make eye contact with the women because we need more of them to come out.

Karina's interest in demonstrating to other women, and especially Latinas, that a career in STEM is reachable goes beyond helping other students; she is also committed to the new generations by trying to reach them. Overall, to try to persuade participants' male peers who often doubt about Latinas' intellectual capacity, participants found ways to stand out, voice their opinion, and especially, fight to be recognized and feel encouraged to accomplish their STEM degree.

## Discussion

The primary research question focuses on *what kind of activities, if any, do Latina college students perform that show leadership styles in STEM-related disciplines?* In this regard, participants' leadership in on-campus activities and even beyond the university setting is outstanding, as described in the findings. Several participants revealed their commitment to improving the college experience of younger students in STEM. Some of them worked as mentors, others were in charge of STEM-related organizations, and some others even participated in outreach activities, trying to influence female high school students to pursue STEM majors. These are but a few of the different activities in which participants are involved and participate as STEM college students. All participants—in some form—were striving to make a change, enhance the college climate, and promote that STEM disciplines can be for all, even women and minorities. In this sense, Eagly and Carli (2007) noted that the social expectation of women leaders is directly related to their gender, including being supportive, helping other individuals to grow, and paying attention to others' needs. The aforementioned statement aligns with the type of activities that participants perform in their role as leaders in STEM.

The first ancillary research question corresponds to *what are the leadership strategies, if any, that Latina college students use to successfully navigate their college experience in male-related disciplines?* While participants perform different activities outside of the classroom that show their leadership, in the classroom they should be able to act and react to face male peers' negative behaviors. Previous research (Banda, 2012; Forster, 2017; Johnson, 2001; Schulze & Tomal, 2006; Smyth & McArdle, 2004) has documented the type of microaggressions, questioning, doubts, confrontation, and problems that women in general, and minoritized women in particular, have to deal with in college. This study corroborated such research, as participants in this study commented on the difficulties they face in classrooms, especially with White male peers.

For the most part, participants were aware of their position in male-dominated disciplines. Some of them mentioned that their involvement in student organizations related to their field allowed them to create a social network, where other women and Latinas support each other. In this sense, the culture of Latinas directly affects their resilience as reflected in the interactions with others, stress management, and sense of community, along with academic and social support. As such, these are all potential factors that influence their resilience (Greene, Galambos, & Lee, 2004). Other participants become more vocal as they proactively voice their opinion in classrooms, making sure everybody knows that women can lead. As a form of human capital, participants develop their own strategies which allow them to advance their education and aspire to continue leading in and out of classrooms (Eagly & Carli, 2007).

The second ancillary research question examines *what type of experiences, if any, influence the leadership adopted by Latina college students who pursue degrees in male-dominated fields?* Participants show their leadership by caring about other students, helping each other, and fighting to be recognized as competent STEM students, creating a better campus-wide ambiance. The experiences of Latinas in this study evidence new ways to define what leadership entails. Aligned with Latinas' culture, their leadership encompasses making alliances with other students, helping others develop academically and emotionally, motivating others to continue their studies or pursue STEM majors, participating more in classes, and working on improving the STEM college climate. Participants' commitment to helping other students succeed contributed to their holistic development and reinforced the ability to manage and conquer the most demanding obstacles through resiliency. Additionally, participants experienced helping others to excel and developed the mentoring ability and affective skills which promoted communal success (Eagly & Carli, 2007). The aforementioned information resonated with several participants, as many discussed taking leadership roles in on-campus organizations and programs as the most appropriate training to become future scientists in leadership positions, which will help Latinas in their future careers.

## Conclusion and Discussion

This study suggested that Latina undergraduate students show important leadership styles as a way to persist in STEM male-dominated disciplines. Some of the leadership skills they perform result from their resilience and ability to become more visible and lift their voices as minoritized students in STEM fields. In this regard, the Leadership Labyrinth Model posits that the intersection between an individual's ethnicity and gender shapes and defines the type of leadership adopted (Eagly & Carli, 2007), and this is especially true for Latinas using their cultural legacy and advocacy for others to lead. In particular, Latinas' leadership proved to be beneficial for fresh-man college students and even high school seniors as Latinas become role models. More importantly, the integration of leadership skills aided Latinas themselves to develop holistically and better prepare for their profession. In addition, Latinas' strong determination, ability to connect with others, and trust in their intellectual capacity helped them overcome barriers and largely influenced their persistence in STEM. The participants of the study dem-onstrated the reinvention of themselves as women, Latinas, and scientists in academic spaces.

Despite literature addressing leadership-related topics of minoritized stu-dents (Beatty, 2015; Guardia, 2015; Kezar & Moriarty, 2000; Lozano, 2015; Onorato & Musoba, 2015) in higher education, there is a research gap on minoritized women, and specifically on how Latinas in STEM disciplines practice leadership. This study contributed to the literature by adding Lati-nas' perspective and highlighting the strategies and activities they adopt to succeed as STEM majors. Such strategies have been useful to stand out, to be known, to be respected and admired inside and outside of the classroom. Latinas recognized that they have the ability to become a force that thrives, surpasses barriers, and finds the necessary tools to lead.

## Implications and Recommendations

A few recommendations for practice include: (1) To reinforce Latinas' empowerment in college, it is important that university departments high-light and promote outstanding Latina women leaders' contributions, not only in leadership programs (Suarez, 2015) but also in workshops and insti-tutional training addressing leadership in college and beyond. (2) To instill in students the importance of acquiring soft skills in college, such as teamwork, leadership, and conflict resolution, STEM faculty should include at least one concept in the form of a class activity in the courses. (3) To develop Latinas' leadership as students and professionals, STEM programs should encourage Latinas to participate in student organizations, especially the organizations related to their college major. Latinas' participation in this type of organization may help develop both leadership attitudes and networking skills.

The gap in the existing leadership literature and the experiences of the participants in this study lend themselves to several recommendations for further studies:

1. *Studying the perception of Latinas' leadership from peers, faculty, and staff perspectives, especially at larger HSIs.* Latinas' leadership can be better comprehended from different lenses, which will help add critical knowledge to the field. In addition, research on this topic could serve to propose programs and services that holistically support the development of Latinas in STEM disciplines in college.
2. *Examining how minoritized women students develop their leadership styles along their college pathway, in trying to find an identity as women, Latinas, and scientists.* Such research could provide meaningful data that help define the safe spaces that Latinas use to excel as STEM college students. This threefold identity (*woman, Latina, and scientist*) could be useful to adopt a leadership style that adapts across the college path.
3. *Analyzing the participation of Latina STEM students to leadership-oriented programs and workshops.* This type of research would contribute to the importance of such programs for Latinas' leadership development. It can also provide evidence of benefits for Latinas who participate in such programs, and through feedback, improve current programs and implement new pilot programs.

## References

Banda, R. M. (2012). *Perceptions of social support networks and climate in the persistence of Latinas pursuing an undergraduate engineering degree* (Doctoral dissertation). Retrieved from ProQuest Dissertations & Theses database. (UMI No. 3537067).

Beatty, C. C. (2015). Latin@ student organizations as pathways to leadership development. In A. Lozano (Ed.), *Latina/o college student leadership: Emerging theory, promising practice* (pp. 65–81). Lanham, MD: Lexington Books.

Cabrera, N. L., & Padilla, A. M. (2004). Entering and succeeding in the "culture of college": The story of two Mexican heritage students. *Hispanic Journal of Behavioral Sciences, 26*(2), 152–170. doi:10.1177/0739986303262604

Eagly, A. H. (2007). Female leadership advantage and disadvantage: Resolving the contradictions. *Psychology of Women Quarterly, 31*(1), 1–12. doi:10.1111/j.1471-6402.2007.00326.x

Eagly, A. H., & Carli, L. L. (2007). *Through the labyrinth: The truth about how women become leaders.* Cambridge, MA: Harvard Business School Press.

Eagly, A. H., & Chin, J. L. (2010). Diversity and leadership in a changing world. *American Psychologist, 65*(3), 216. doi:10.1037/a0018957

Erlandson, D. A., Harris, E. L., Skipper, B. L., & Allen, S. D. (1993). *Doing naturalistic inquiry: A guide to methods*. Newbury Park, CA: Sage.

Feagin, J. R. (2013). *The White racial frame: Centuries of racial framing and counter-framing*. New York, NY: Routledge.

Forster, E. (2017, May 4). *As a woman in science, I need to conceal my femininity to be taken seriously*. Retrieved from Vox Media website: www.vox.com/first-person/2017/5/4/15536932/women-stem-science-feminism

Foulis, E. (2017). Latina/o college student leadership: Emerging theory, promising practice. *Journal of Latinos and Education, 16*(2), 167–168. doi:10.1080/15348431.2016.1205994

Glesne, C. (2011). *Becoming a qualitative researcher* (4th ed.). Boston, MA: Pearson.

Greene, R. R. (2002). *Resilience: Theory and research for social work practice*. Washington, DC: NASW Press.

Greene, R. R., Galambos, C., & Lee, Y. (2004). Resilience theory: Theoretical and professional conceptualizations. *Journal of Human Behavior in the Social Environment, 8*(4), 75–91. doi:10.1300/j137v08n04_05

Guardia, J. R. (2015). Leadership and identity. In A. Lozano (Ed.), *Latina/o college student leadership: Emerging theory, promising practice* (pp. 3–28). Lanham, MD: Lexington Books.

Haber-Curran, P., Miguel, M., Shankman, M. L., & Allen, S. (2018). College women's leadership self-efficacy: An examination through the framework of emotionally intelligent leadership. *NASPA Journal About Women in Higher Education, 11*(3), 297–312. doi:10.1080/19407882.2018.1441032

Johnson, A. (2001). Women, race, and science: The academic experiences of twenty women of color with a passion for science. *Dissertation Abstracts International, 62*(02), 428.

Jordan, J. V., Hartling, L. M., & Walker, M. (Eds.). (2004). *The complexity of connection: Writings from the Stone Center's Jean Baker miller training institute*. New York, NY: Guilford Press.

Kark, R., & Eagly, A. H. (2010). Gender and leadership: Negotiating the labyrinth. In J. Chrisler & D. McCreary (Eds.), *Handbook of gender research in psychology* (pp. 443–468). New York, NY: Springer.

Kendricks, K. D., Arment, A. A., Nedunuri, K. V., & Lowell, C. A. (2019). Aligning best practices in student success and career preparedness: An exploratory study to establish pathways to STEM careers for undergraduate minority students. *Journal of Research in Technical Careers, 3*(1), 27. doi:10.9741/2578-2118.1034

Kezar, A., & Moriarty, D. (2000). Expanding our understanding of student leadership development: A study exploring gender and ethnic identity. *Journal of College Student Development, 41*(1), 55–69.

Lincoln, Y., & Guba, E. (1985). *Naturalistic inquiry*. Thousand Oaks, CA: Sage.

Lozano, A. (2015). Re-imaging Latina/o student success at a historically White institution. In A. Lozano (Ed.), *Latina/o college student leadership: Emerging theory, promising practice* (pp. 3–28). Lanham, MD: Lexington Books.

Miller, J. B. (1986). *Toward a new psychology of women*. Boston, MA: Beacon Press.

Mitts, C. R. (2016). Why STEM? *Technology & Engineering Teacher, 75*(6), 30–35.

National Science Foundation (2016). Chapter 2: Higher education in science and engineering. In *Science & engineering indicators*. Retrieved from https://www.nsf.gov/statistics/2016/nsb20161/uploads/1/12/chapter-2.pdf

National Science Foundation (2017). *Women, minorities, and persons with disabilities in science and engineering: 2017.* Retrieved from https://www.nsf.gov/statistics/2017/nsf17310/data.cfm

Onorato, S., & Musoba, G. D. (2015). La líder: Developing a leadership identity as a Hispanic woman at a Hispanic-serving institution. *Journal of College Student Development, 56*(1), 15–31. doi:10.1353/csd.2015.0003

Schoon, I. (2015). Explaining persisting gender inequalities in aspirations and attainment: An integrative developmental approach. *International Journal of Gender, Science, and Technology,* 7(2), 151–165.

Schulze, E., & Tomal, A. (2006). The chilly classroom: Beyond gender. *College Teaching, 54*(3), 263–269. doi:10.3200/ctch.54.3.263-270

Smyth, F. L., & McArdle, J. J. (2004). Ethnic and gender differences in science graduation at selective colleges with implications for admission policy and college choice. *Research in Higher Education, 45,* 353–381. doi:10.1023/b:rihe.0000027391.05986.79

Suarez, C. E. (2015). Never created with nosotros in mind. In A. Lozano (Ed.), *Latina/o college student leadership: Emerging theory, promising practice* (pp. 29–43). Lanham, MD: Lexington Books.

Van Breda, A. D. (2001). *Resilience theory: A literature review.* Pretoria, South Africa: South African Military Health Service.

Yin, R. K. (2009). *Case study research: Design and methods* (4th ed.). Thousand Oaks, CA: Sage.

# 11 "There Was Something Missing"

## How Latinas Construct Compartmentalized Identities in STEM

*Ariana L. Garcia, Blanca Rincón, and Juanita K. Hinojosa*

> [W]hen you are a minority, especially pre-med, and as you go up higher in the classes, you see less of yourself, and you already know there is competition ahead of you. It is nothing new, it is just ingrained in me. I have to compete against myself and know there are less people like me, and I know some people might say things—like a classic thing some people might say is she got this position [because] she is Hispanic. [. . .] People might not say it, but they think about it.
>
> —Juana, health and biological sciences major

Despite significant gains in Latinas' access to postsecondary education, gaps remain for Latinas pursuing degrees in the fields of science, technology, engineering, and mathematics (STEM). In 2017, Latinas received one in five of all undergraduate engineering degrees awarded to Latinxs in the United States (National Science Board, 2018). Most STEM jobs require some form of postsecondary education (e.g., certificate, degree); consequently, it is not surprising that Latinas comprise less than 2% of the nation's science and engineering workforce (National Science Foundation, 2015). Subsequently, Latinas' limited access to STEM credentials has significant consequences for our nation's STEM labor pool.

Several researchers have investigated Latina underrepresentation in the STEM pipeline (Banda & Flowers, 2018; Leyva, 2016; Rincón & Lane, 2017; Rodriguez, Doran, Sissel, & Estes, 2019; Villa & González, 2014). Latinas, like other Women of Color, experience STEM contexts through multiple markers of marginalization due to their gender and racial/ethnic identities. Scholars describe living at the intersection of multiple minoritized statuses as experiencing a "double bind" (Ong, Wright, Espinosa, & Orfield, 2011). As for women of color, these experiences are often marked by a series of risk factors within the context of STEM, including experiencing isolation (Brown, 2008), alienation, and covert and overt forms of racism and sexism (Garriott et al., 2019). In the epigraph that begins this chapter, Juana describes how this marginalization is ubiquitous and further intensifies as Latinas make progress towards upper-division STEM course-work and beyond.

The emotional and psychological consequences for Latinas and other women of color, who constantly negotiate instances of racism and sexism in STEM environments, are well documented (Banda & Flowers, 2018; Gibson & Espino, 2016). Women employ a variety of coping strategies in response to these risk factors, including exercising agency (Ko, Kachchaf, Hodari, & Ong, 2014), self-isolation (Ko et al., 2014), and censoring or distancing themselves from their marginalized identities (Banda & Flowers, 2018; Gibson & Espino, 2016; Leyva, 2016). In fact, women construct a wide range of identities to navigate STEM environments: multidimensional, compartmentalized, and isolated identities. Women with multidimensional identities describe their identities as complex and intersectional (Cegile, 2011; Rodriguez, Friedensen, Marron, & Bartlett, 2019; Verdin & Godwin, 2018). Other women recount living fragmented lives where they compartmentalize aspects of their identities depending on the context (Gibson & Espino, 2016). Still others (mainly Latinas) construct isolated identities where their most salient identity is foregrounded (Banda & Flowers, 2018; Verdin & Godwin, 2018). How Latinas and other women of color make meaning of their identities has important implications for their persistence and success in STEM.

## Resilience and Identity

While the extant literature focuses on how the social markers of Latinas and other women of color present unique "risk factors" in STEM that need to be overcome, more recently scholars have highlighted the ways in which social group membership may provide culturally specific forms of support and resources that minimize the risks that students encounter in STEM contexts (McGee & Spencer, 2012). In this way, students exhibit resilience when they draw on "protective processes (resources, competencies, talents, and skills) that sit within the individual" (Olsson, Bond, Burns, Vella-Brodrick, & Sawyer, 2003, p. 3). For example, McGee and Spencer (2012) describe how racial group membership (or racial identity) embodies the protective factors of family, group solidarity, and cultural knowledge. Similarly, Esteban-Guitart and Moll (2014) argue that identity reflects the culturally developed "funds of knowledge" that one accumulates through their lived experience. These funds of knowledge encompass all the "people, skills, knowledge, practices, and resources that people have acquired and now use through their involvement in various activities" (Esteban-Guitart & Moll, 2014, p. 37). In this way, we understand students' social identities as a protective factor that facilitates resilience and leads to persistence in STEM.

# Methods

The current study is informed by data gathered from a large, multi-year, complementarity sequential mixed-methods study (Greene, Caracelli, & Graham, 1989) funded by the National Science Foundation. The mixed-methods design for the larger study uses both quantitative and qualitative data to examine facets of the college experience and career decision-making processes for Students of Color in STEM fields (Greene et al., 1989). While the larger study aims to longitudinally investigate how Students of Color pursue STEM degrees, this sub-study aims to understand how Latinas pursuing STEM degrees understand and utilize their identities to persist in STEM. We specifically examine the following research questions: *How do Latinas pursuing STEM degrees describe their identities? How do Latinas draw on their identities to persist in STEM?*

### Researchers' Positionalities

For any qualitative study, it is important to discuss how the social position of the researcher influences the research process, including data collection, analysis, and interpretations (Creswell, 2013; Saldaña, 2018). This research study was conducted by three Latinas, two of whom are first-generation college students. One researcher is a professor in the school of education at a large public university, and the other two researchers are graduate students who are pursuing their doctorates in education. Our scholarly interests broadly focus on equity issues for traditionally underserved students in higher education.

As researchers, we believe that our lived experiences as Latinas who were primarily the first in their family to go to college, were assets that we could leverage during the research process. As such, we drew upon our collective cultural intuition (Delgado Bernal, 1998). That is, we used our "personal experience, collective experience, professional experience, [and] communal memory" (Delgado Bernal, 2016, p. 1) to gain a unique insight and understanding into how participants described and experienced their identities, and how these identities were engaged to persist in STEM (Leedy & Ormrod, 2005).

### Data Sources and Participants

Study participants included first-time, full-time, first-year, and transfer students matriculating at one of four universities in the Northeastern United States in the fall of 2016. The four institutions represent a variety of institutional contexts, including large public land-grant universities, a large private urban university, and a STEM-focused private urban university. While study participants attended a variety of institutions, they all participated in

*Table 11.1* Latina Student Profiles (*n*=8).

| Pseudonym | Race/Ethnicity | Major | First-Generation Status |
| --- | --- | --- | --- |
| Josie | Latinx/White | Engineering | No |
| Luisana | Latinx | Engineering | Yes |
| Jazmine | Latinx | Engineering | Yes |
| Jordan | Latinx/White | Engineering | No |
| Marisela | Black/Latinx | Engineering | No |
| Erica | Latinx | Biology & Health Sciences | No |
| Awilda | Black/Latinx | Biology & Health Sciences | Yes |
| Juana | Latinx | Biology & Health Sciences | Yes |

the Louis Stokes Alliance for Minority Participation (LSAMP) program, a federally funded program that aims to increase the number of underrepresented minority students matriculating into, and successfully completing, high-quality undergraduate degrees in STEM.

Data collection for this study occurred over several academic school years. During the 2016–2017 academic year, members of the research team traveled to each campus and invited students to participate in the study. Study participants completed an in-person survey that gathered their demographic information, including race/ethnicity, gender, and first-generation college status (see Table 11.1). In the second year of the study, students received an email asking them to participate in a follow-up one-on-one interview. A total of eight self-identified Latina students agreed to participate. Among them, half identified with two ethnic or racial categories, representative of the complex and multidimensional aspects of Latina identity.

The interviews were conducted using a semi-structured protocol that included questions about students' involvement on campus and who they went to when they needed advice about their future in STEM. Before conducting the interviews, members of the research team tested and received feedback on the protocol to ensure that the questioning route was conversational and that the questions were relevant, clear, and concise (Krueger & Casey, 2009). All interviews were conducted in person and took place in a mutually agreed upon space on the student's college campus. Each interview was audio recorded with the consent of participants and ranged between approximately 40 and 80 minutes in length.

### Data Analysis

To analyze the qualitative data, the research team transcribed each audio file and reviewed each transcript for accuracy. Then, we de-identified the transcripts by replacing students' names and other identifying information with pseudonyms. Next, we entered the interview transcript data and open-ended survey responses into Dedoose, an online software used for analyzing qualitative data. To begin, two members of the research team used open coding techniques (Saldaña, 2015) to develop an initial codebook—that is, a shared understanding of data concepts and themes that would inform future rounds of coding (Denzin & Lincoln, 2008).

Once we defined an initial set of codes, two members of the research team independently and carefully read and analyzed the interview transcripts. Then, we came together to identify similarities and differences in our coding of each interview transcript until discrepancies were resolved. When coding discrepancies were encountered, a research team member served as a peer-debriefer (Saldaña, 2015). Reconciling discrepancies among researchers strengthened the codes and increased the reliability of the analysis (Miles, Huberman, & Saldaña, 2014). As a multi-member team, peer debriefing and memoing served as critical components in the data analysis process (Saldaña, 2015). Finally, members of the research team engaged in a second round of coding to develop themes.

## Findings

Two major themes emerged from our analysis of the data. First, we found that most Latinas in STEM described their identities as compartmentalized. Second, Latinas leveraged these compartmentalized identities to establish same-identity mentoring relationships with others in STEM who nurtured their identity development and supported their persistence. We describe each theme further in the following sections.

### Constructing a Compartmentalized Identity

Latinas often compartmentalized aspects of their identities and utilized different involvement opportunities to nurture specific facets of their identities. For example, some participants focused on becoming involved in student organizations that nurtured their racial/ethnic identity, while others became involved in co-curricular experiences that nurtured their science, gender, religious, or sexual identities—highlighting the complex identities of Latinas pursuing STEM degrees.

Erica, a biology and health sciences major, shared the need to have "a broad range of friends [. . .] in every aspect of [her] life." Erica described becoming very involved with several STEM-focused student organizations and learning communities upon first arriving at her institution but feeling like "there was something missing" as she progressed to the end of her second year in college. Because she grew up in a predominantly White and rural town, she wanted to explore her racial/ethnic heritage during her college years. She did so through her involvement in a Latina sorority and participating in her institution's Latinx Cultural Center. By developing these separate friend groups to meet her academic and cultural needs, Erica implicitly speaks to the lack of diversity found within STEM fields. Because there were few places within her institution where she could simultaneously attend to her racial/ethnic and STEM identities, she was forced to compart-mentalize these aspects of her identity.

How students compartmentalized their identities influenced the college activities they engaged in. Josie described how she decided not to become involved with the Society for Hispanic Professional Engineers (SHPE) after attending a meeting. She explained:

> [T]here was nothing wrong with [SHPE]. I think I just thought I didn't really need that community and I was going to use the [engineering student organization] to sort of be my professional development and then I was going to connect with my Hispanic or Spanish part through my minor so I just didn't want another obligation that was going to be repetitive I guess.

Because Josie utilized her Spanish minor as a way to connect with and nur-ture her cultural identity, she did not "need" SHPE as a community. In fact, Josie described each of her co-curriculars as intentionally and separately attending to various aspects of her identity: engineer and race/ethnicity.

Unlike Josie and Erica, Awilda described spaces within STEM where she attended to multiple aspects of her identity. For example, Awilda became involved with LSAMP to have "another support group [. . .] because [she was] out here alone practically." Awilda recalled Tonisha, her advisor, say "you're Latina, Black, first-generation student, you're out of state, I'm going to introduce you to the LSAMP program." This further highlights that when there are places for Latinas to connect with multiple aspects of their identities within STEM, these opportunities were beneficial for students.

*Identity and Persistence*

Because Latinas in this study learned to compartmentalize aspects of their identities, they were able to utilize these compartmentalized identities to develop same-identity mentoring relationships that nurtured their identity development and supported their persistence. As such, these women devel-oped close relationships with mentors in STEM who shared their background as either a woman, a first-generation college attendee, or a member of a racial/ethnic minoritized group.

Awilda began working in Dr. Everette's research lab as a first-year stu-dent. She was drawn to Dr. Everette because of their shared research inter-ests and backgrounds as the first in their families to pursue postsecondary education. Having someone who understood what she was going through without having to explain it was invaluable to Awilda. She shared, it "gives me comfort because I can talk to them about things that my friends will not understand, and they see my struggle in school, especially with confidence."

Because feelings of self-doubt are prevalent amongst first-generation college students, her mentor, having shared similar experiences, actively validated Awilda's contributions to his research lab. She reflected on how the compliments she received from her mentor "[meant] a lot" and showed "he really does care," especially when her achievements were shared with other professors. In this way, her mentor served as both a critical source of external validation of her STEM competencies and as an advocate that communicated her accomplishments with others who could potentially reinforce her STEM talents and abilities.

Similar to Awilda, Juana's mentor, Mary, was also the first in her family to go to college. In describing Mary's background, she explained, "She is from Cali, I think she grew up in the 'hood. So, yeah, she gets it." Mary was a constant source of support for Juana, who described Mary as her "rock". In explaining their relationship, Juana shared that Mary was "one of those people I can just call for everything. [. . .] [I]f I am having a bad day, she will be there for me." As Juana reflected, Mary "gets her" and was able to support her academically and personally. Here we see how Juana and Mary's shared background provided a mutual understanding of the psychosocial support that first-generation college students need to persist in higher education—a holistic approach that goes beyond academics.

Other students found mentorship through their university-sponsored co-op. Josie's mentor, Carol, whom she described as having a "very similar background" as her, had attended the same university, had the same major, shared the same co-op experience, and identified as a woman. It was this shared experience that made it easy for Josie to seek advice from Carol about her future in engineering. For example, Josie turned to Carol to navigate a difficult situation with her co-op supervisor.

> So, I sort of talk to her because she has had him as a boss, and I was like, how can I make him feel more comfortable letting me do this? How can I improve these particular things? And so, she sort of helped me separate between things that I can improve, things that were stylistic differences between us as teachers, and ways to make him feel more comfortable with me teaching the material and know that I was very confident and I was going to do as good of a job as I could.

In a field historically dominated by men, Josie struggled to get her supervisor to trust her in performing specific tasks. Because of their shared gender identity, Josie was able to turn to Carol for help in order to navigate the sexism that is often pervasive within engineering. Carol, who had experienced a similar situation with her supervisor, was able to give her advice on this situation and other engineering-related concerns. Josie was grateful to be able to turn to her mentor for advice on how to assert herself as knowledgeable and capable.

## Discussion

The findings from this study offer new perspectives on how Latinas make meaning of their identities within STEM, and how their constructed identities aid in their subsequent persistence. First, our findings challenge conventional understandings of how Latinas make meaning of their identities within STEM contexts. To date, researchers have consistently found that Latinas in STEM are more apt to construct isolated identities that privilege their gender identity (Banda & Flowers, 2018; Brown, 2008). Our findings contradict prior research by demonstrating how some Latinas experience fractured and compartmentalized identities similar to Black women in STEM (Gibson & Espino, 2016). When opportunities were available that attended to multiple aspects of their identities, participants engaged and benefited from these opportunities. However, when these needs were unmet, students had to locate other spaces within the university to attend to and nurture these aspects of their identities. For example, to connect to her racial/ethnic identity, Erica turned to her university's Latinx Cultural Center. In doing so, attending to her racial/ethnic identity became an additional "obligation" on top of other time-intensive commitments for her major. At the same time, this nurturing of her racial/ethnic identity became central to developing a support system on campus that aided in her persistence.

To this point, previous research has found that Latinx students who actively engage in diversity-related activities outside of STEM contexts have lower GPAs and are less likely to persist to degree completion (Cole & Espinoza, 2008). Because some students have to go outside of STEM to find cultural support, these activities can unintentionally serve as mechanisms that "pull" Latinas away from STEM. Further research is needed to explore how identity compartmentalization is related to institutional shortcomings—that is, the lack of involvement opportunities on campus that attend to Latinas' holistic and intersectional identities. Expanding on this research would also provide an opportunity to better understand whether/how programs or organizations that are intersectional (e.g., SHPE señoritas) support Latinas and other women of color in constructing multidimensional understandings of their identities within STEM.

Second, this study finds that students' social identities serve as a protective factor that facilitates resilience and leads to student persistence in STEM in two important ways. First, Latinas' involvement in identity-specific organizations and same-identity mentoring relationships nurtured different aspects of their identities and provided an important source of support and belonging. However, because most participants were unable to find spaces within STEM that attended to their complex and multidimensional identities, Latinas showed resilience in their ability to adapt and locate university spaces (in and outside of STEM) that attended to these aspects of their identities. In this way, participants' construction of compartmentalized identities demonstrates the resilience that is required of students to persist in STEM, further reinforcing how identity is used as a protective factor to negate risk factors (e.g., isolation, alienation, tokenization) within STEM. While this tactic may alleviate immediate risks that arise in STEM, little is known about the long-term consequences for Latinas and other women of color, who constantly negotiate their identities as they enter new spaces. This raises the question, what are the long-term consequences of utilizing this strategy for Latinas and other women of color?

Finally, findings from this study also support the important role that identity plays in the development of mentoring relationships that positively influence Latinas' persistence in STEM. While research demonstrates that Latinx students actively prefer and seek out same-race mentors (Medina & Posadas, 2012), none of the Latinas in this study identified having a Latinx mentor. It was through their construction of identities compartmentalized identities, then, that Latinas were able to find and cultivate same-identity relationships within a context that lacks both racial/ethnic and gender diversity. These mentoring relationships provided students access to resources, opportunities, and psychosocial support—buffering the risk factors that students encounter as they navigate higher education and STEM (Griffin et al., 2018). Further, these mentors served as a critical source of validation of their STEM talents and abilities. Because Latinas are largely dependent on the recognition of others to view themselves as scientists or engineers (Cegile, 2011; Rodriguez, Cunningham, & Jordan, 2019), the recognition that these students received from their mentors aided their development of a STEM identity—a factor that has been positively associated with success in STEM (Rodriguez et al., 2019). It is worth noting that Latinas and other women of color often find it difficult to cultivate a STEM identity—this is because their gender, race, and social class often serve as barriers to being recognized by others (Carlone & Johnson, 2007; Cegile, 2011; Gibson & Espino, 2016; Rodriguez et al., 2019). This finding shines light on the need to investigate the relationship between same-identity mentorship and STEM identity development for Latinas.

## Conclusion

As Latinxs are the fastest-growing ethnic group in the United States, higher-education institutions must be strategic about enrolling them and establishing initiatives that support their holistic development. The Latinas in our study exemplified their resilience by overcoming the challenges they faced within STEM, particularly by nurturing and compartmentalizing their identities. And while we should recognize and validate Latinas' ability to thrive in spaces that were not created with them in mind, we must further analyze the systemic inequities embedded within higher education that require Latinas to be resilient in the first place. To this end, institutions should assess the availability of programs that support students' holistic identity development. Further, institutions should support existing efforts that intentionally connect Latinas to opportunities where they can form mentoring relationships with other Latinxs who can provide advice, support, and guidance that is responsive to their multidimensional identities (e.g., LSAMP). If institutions are serious about making strides towards increasing Latina participation and success in STEM, these institutional shortcomings must be addressed.

## References

Banda, R. M., & Flowers, A. M. (2018). Critical qualitative research as a means to advocate for Latinas in STEM. *International Journal of Qualitative Studies in Education, 31*(8), 769–783. doi:10.1080/09518398.2018.1479046

Brown, S. (2008). The gender differences: Hispanic females and males majoring in science and engineering. *Journal of Women and Minorities in Science and Engineering, 14*(2), 205–223. doi:10.1615/jwomenminorscieneng.v14.i2.50

Carlone, H. B., & Johnson, A. (2007). Understanding the science experiences of successful women of color: Science identity as an analytic lens. *Journal of Research in Science Teaching, 44*(8), 1011–1245. doi:10.1002/tea.20237

Cegile, R. (2011). Underrepresentation of women of color in the science pipeline: The construction of science identities. *Journal of Women and Minorities in Science and Engineering, 17*(3), 271–293. doi:10.1615/jwomenminorscieneng.2011003010

Cole, D., & Espinoza, A. (2008). Examining the academic success of Latino students in science technology engineering and mathematics (STEM) majors. *Journal of College Student Development, 49*(4), 285–300. doi:10.1353/csd.0.0018

Creswell, J. W. (2013). *Qualitative inquiry and research design: Choosing among five approaches.* Thousand Oaks, CA: Sage.

Delgado Bernal, D. (1998). Using a Chicana feminist epistemology in educational research. *Harvard Educational Review, 68*(4), 555–582. doi:10.17763/haer.68.4.5wv 1034973g22q48

Delgado Bernal, D. (2016, June). *Cultural intuition: Then, now, and into the future* (Research Brief No. 1). Retrieved from Center for Critical Race Studies at UCLA website: www.ccrse.gseis.ucla.edu/publications

Denzin, N. K., & Lincoln, Y. S. (2008). *Strategies of qualitative inquiry* (3rd ed.). Thousand Oaks, CA: Sage.

Esteban-Guitart, M., & Moll, L. C. (2014). Funds of Identity: A new concept based on the funds of knowledge approach. *Culture & Psychology, 20*(1), 31–48. https://doi.org/10.1177/1354067X13515934

Garriott, P. O., Navarro, R. L., Flores, L. Y., Lee, H. S., Carrero Pinedo, A., Slivensky, D., . . . Luna, L. (2019). Surviving and thriving: Voices of Latina/o engineering students at a Hispanic serving institution. *Journal of Counseling Psychology, 66*(4), 437–448. doi:10.1037/cou0000351

Gibson, S. L., & Espino, M. M. (2016). Uncovering Black womanhood in engineering. *NASPA Journal About Women in Higher Education, 9*(1), 56–73. doi:10.1080/1940788 2.2016.1143377

Greene, J. C., Caracelli, V. J., & Graham, W. F. (1989). Toward a conceptual framework for mixed-method evaluation designs. *Educational Evaluation and Policy Analysis, 11*(3), 255–274. doi:10.3102/01623737011003255

Griffin, K., Baker, V., O'Meara, K., Nyunt, N., Robinson, T., & Staples, C. L. (2018). Supporting scientists from underrepresented minority backgrounds: Mapping developmental networks. *Studies in Graduate and Postdoctoral Education, 9*(1), 19–37. doi:10.1108/sgpe-d-17–00032

Ko, L. T., Kachchaf, R. R., Hodari, A. K., & Ong, M. (2014). Agency of women of color in physics and astronomy: Strategies for persistence and success. *Journal of Women and Minorities in Science and Engineering, 20*(2), 171–195. doi:10.1615/jwomenminorscieneng.2014008198

Krueger, R. A., & Casey, M. A. (2009). *Focus groups: A practical guide for applied research.* Thousand Oaks, CA: Sage.

Leedy, P. D., & Ormrod, J. E. (2005). *Practical research: Planning and design* (8th ed.). Upper Saddle River, NJ: Pearson Hall.

Leyva, L. A. (2016). An intersectional analysis of Latin@ college women's counter-stories in mathematics. *Journal of Urban Mathematics Education, 9*(2), 81–121.

McGee, E., & Spencer, M. B. (2012). Theoretical analysis of resilience and identity. In E. Dixon-Román & E. W. Gordon (Eds.), *Thinking comprehensively about education: Spaces of educative possibility and their implications for public policy* (pp. 161–178). London: Routledge.

Medina, C. A., & Posadas, C. E. (2012). Hispanic student experiences at a Hispanic-serving institution: Strong voices, key message. *Journal of Latinos and Education, 11*, 182–188. doi:10.1080/15348431.2012.686358

Miles, M. G., Huberman, A. M., & Saldaña, J. (2014). *Qualitative data analysis: A methods sourcebook* (3rd ed.). Los Angeles, CA: Sage.

National Science Board. (2018). *Science and engineering indicators 2018* (Publication No. NSB-2018–1). Retrieved from National Science Foundation website: www.nsf.gov/statistics/indicators/

National Science Foundation (2015). *Women, minorities, and persons with disabilities in science and engineering: Overall trends* (Special Report NSF 15–311). Retrieved from National Science Foundation website: www.nsf.gov/statistics/2017/nsf17310/digest/occupation/overall.cfm

Olsson, C. A., Bond, L., Burns, J. M., Vella-Brodrick, D. A., & Sawyer, S. M. (2003). Adolescent resilience: A concept analysis. *Journal of Adolescence, 26*(1), 1–11. doi:10.1016/s0140–1971(02)00118–5

Ong, M., Wright, C., Espinosa, L. L., & Orfield, G. (2011). Inside the double bind: A synthesis of empirical research on undergraduate and graduate women of color in science, technology, engineering, and mathematics. *Harvard Educational Review, 81*(2), 172–208. doi:10.17763/haer.81.2.t022245n7x4752v2

Rincón, B., & Lane, T. B. (2017). Latin@s in science, technology, engineering, and mathematics (STEM) at the intersections. *Journal of Equity and Excellence in Education, 50*(2), 182–195. doi:10.1080/10665684.2017.1301838

Rodriguez, S. L., Cunningham, K., & Jordan, A. (2019). STEM identity development for Latinas: The role of self- and outside recognition. *Journal of Hispanic Higher Education, 18*(3), 254–272. doi:10.1177/1538192717739958

Rodriguez, S. L., Doran, E. E., Sissel, M., & Estes, N. (2019). Becoming la ingeniera: Examining the engineering identity development of undergraduate Latina students. *Journal of Latinos and Education, 1–20.* doi:10.1080/15348431.2019.1648269

Rodriguez, S. L., Friedensen, R., Marron, T., & Bartlett, M. (2019). Latina undergraduate students in STEM: The role of religious beliefs and STEM identity. *Journal of College and Character, 20*(1), 25–46. doi:10.1080/2194587x.2018.1559198

Saldaña, J. (2015). *The coding manual for qualitative researchers* (3rd ed.). Los Angeles, CA: Sage.

Saldaña, J. (2018). Researcher, analyze thyself. *The Qualitative Report, 23*(9), 2036–2046. doi:10.1177/1609406918801717

Verdin, D., & Godwin, A. (2018). Exploring Latina first-generation college students' multiple identities, self-efficacy, and institutional integration to inform achievement in engineering. *Journal of Women and Minorities in Science and Engineering, 24*(3), 261–290. doi:10.1615/jwomenminorscieneng.2018018667

Villa, C., & González, E. (2014). Women students in engineering in Mexico: Exploring responses to gender differences. *International Journal of Qualitative Studies in Education, 27*(8), 1044–1061. doi:10.1080/09518398.2014.924636

# Afterword

## Six Steps Forward for Studying Diversity, Equity, and Inclusion in STEM

*Frank Fernandez*

We worked to compile a volume that offers examples of how scholars study resilience. Now, to conclude the volume, I will reflect on some of the lessons from contributors and recommend six steps forward for scholars who seek to study diversity, equity, and inclusion in STEM. Although I hope to broaden directions for future research to include multiple groups of students, I see these steps as essential for continuing to advance the literature on Latinas in STEM.

Now more than ever, the nation's economic and scientific competitiveness depends on expanding opportunities for Latinas to succeed in science. Improving equity in STEM fields is imperative if institutions of higher education in the United States wish to live up to their rhetoric about equal opportunity. Over the last century, women's access to higher education expanded so rapidly that women outnumbered men among associate's-degree earners by 1977, bachelor's-degree earners by 1981, master's-degree earners by 1986, and doctorate-degree earners by 2005 (Snyder & Dillow, 2012, Table 310). Yet, women's degree attainment continues to vary across fields and remains persistently low in STEM. I aim to identify research directions so that scholars and practitioners can close the opportunity gap and achieve equitable levels of degree attainment for women and minorities in STEM.

First, I consider that several chapters in the volume refer to the intersectionality of gender and race and argue that scholars should continue to explore those connections. Then, I suggest that scholars also acknowledge that individuals' gender and racial identities may change over time. After that, I posit that scholars may be able to get additional traction for understanding STEM education and advocating for promising practices by studying broad groups of students. My fourth point speaks to the need to do international and comparative research on STEM education. Fifth, I claim that scholars should expand their methodological approaches to studying STEM by including critical quantitative techniques. Finally, I return to the volume's central theme—resilience—and consider how scholars may apply the term in future discourse. I see these six directions as promising avenues for understanding broader equity challenges in higher education.

## Untangle the Double Bind: Analyze the Intersectionality of Gender and Race

Drawing on previous work (Malcom, Hall, & Brown, 1976; Ong, Wright, Espinosa, & Orfield, 2011) and the concept of intersectionality (e.g., Crenshaw, 1989), several contributors to this volume mention that Latinas in STEM experience a "double bind" and are minoritized based on both their gender and race (e.g., Kim, Beverly, & Ro, 2020; Mein, Guerra, & Herrera-Rocha, 2020; Garcia, Rincón, & Hinojosa, 2020). Grimes and Morris (1997) recount a Latina sociologist stating that her gender is far more influential in her experiences in academia than her racial identity; she also states that her working-class background is more important than her racial identity but less salient than her gender identity. Moving forward, scholars should continue to analyze the ways in which both gender and race influence how students navigate STEM. In other words, studies should examine where, when, and how the two identities matter.

Consider two undergraduate Latinas: one studies biology and the other studies physics. To what extent is the experience in biology—where women outnumber men (National Science Foundation, 2017)—influenced by being a member of a racial minority group as opposed to a member of a gender majority group? How is that different from the experience of a Latina in physics, where both identities are minoritized? Literature suggests that the gender bond of the double bind may be more salient for a Latina in physics than in biology. For example, Gonsalves (2011) conclude that although "the culture of physics was espoused as gender neutral . . . gender is woven into the cultural narratives of disciplines like physics in ways that students do not detect, yet position themselves around nonetheless" (p. 130). We must better understand double-bind situations so that we may work to unravel them.

## Consider How Gender and Racial Identities Change Over Time

The STEM pathway begins in elementary school, where girls receive early preparation for math and science. Later, young women in high school make decisions about where to apply to college and choose initial academic majors. Women apply to graduate school toward the end of, or after completing, undergraduate work. By the time women pursue faculty or postdoctoral careers in STEM, at least two decades have passed since they were introduced to math and science.

In her study of graduate school choice, Kallio (1995) argues that we need to consider factors (e.g., marriage, having children) that lead to life-stage differences. For example, motherhood typically changes how women understand their gender identity based on their roles as mothers (McMahon, 1995). According to Grimes and Morris (1997), women may be more likely to favor family commitments and make educational or career sacrifices. If we are to improve women's experiences in STEM as undergraduates, graduate students, faculty, or other professionals, then we need to further examine life-stage differences.

Just as women may reconsider their gender roles over time and across life stages, college students' racial identities may evolve based on individuals' situations (Renn, 2000). Fuller-Rowell, Burrow, and Ong (2011) explain that there are three dimensions to racial identity. *Centrality* refers to how an individual identifies with a racial group. *Private regard* describes whether an individual feels positively or negatively about a racial group—and being part of that group. Compared to private regard, *public regard* has to do with an individual's perceptions about whether others feel positively or negatively about a racial minority group (Fuller-Rowell et al., 2011). Renn (2000) finds that demographic factors and peer culture in college contexts can influence how multiracial students make sense of their racial identities.

Additionally, students may have "encounter" experiences that lead them to explore their racial identities. An encounter experience is a "significant personal or social race-related event that is inconsistent with an individual's existing frame of reference," which challenges a person "to think through their existing attitudes and beliefs and to consider various other possible perspectives relating to their race" (Fuller-Rowell et al., 2011, p. 1609; see also Cross, 1991). Although Renn (2000) focuses on situational factors on a college campus, Fuller-Rowell and colleagues (2011) shows that factors outside the immediate college context—for example, a national presidential election—can be meaningful encounters that spur students to explore their racial identities. When Barack Obama was elected President of the United States, Black undergraduates experienced increases in multiple dimensions of racial identity (Fuller-Rowell et al., 2011). One avenue for future research and practice is to classify different types of encounters that trigger positive and negative racial identity exploration among racially minoritized individuals throughout STEM education and careers.

## Not Just Women, Not Just Latinas: Studying Phenomena and Not Groups

In addition to studying disparate groups or specific practices, which we have done in this volume, we need to consider how STEM communities are created and maintained—and to what extent they are inclusive. Scholars may study social phenomena like the distribution of power or what makes STEM academic and professional communities possible. While we are interested in the numerical representation of gendered and racially minoritized groups within STEM, we are not just interested in understanding individual characteristics or experiences. It *is* helpful to have counter-narratives that remind us not to adopt a deficit approach when thinking about student attrition within STEM. Beyond that, we are interested in improving social institutions. While keeping students central in our work, we need to examine power dynamics within universities, academic departments, and research teams.

This volume addresses the underrepresentation of Latina women in STEM. Yet, if we want to understand the broader challenges impeding diversity, equity, and inclusion in STEM, then we need to also study men and other racial minority groups. It is too easy to interpret findings about Latinas as specific characteristics of a single group. In fact, this is good research training. Researchers should not carelessly generalize.

However, studies of multiple groups of students lead to similar findings as the chapters in this volume. For example, multiple authors in this volume highlight the importance of mentoring Latinas (Garcia et al., 2020; Gonzalez, Molina, & Turner, 2020), which mirrors earlier work that examines Black and Latino men (e.g., Strayhorn, Long, Kitchen, Williams, & Stenz, 2013). Therefore, one way to advance future research is to focus on developing a research agenda around mentoring in STEM that includes multiple student groups, rather than to have disparate studies for combinations of minoritized statuses. Studying phenomena and not individual groups may even help researchers advocate for solutions that are expected to benefit more than a single group. How does decision-making work when it comes to choosing research topics, use of lab equipment, and the order of authors on publications? How could it work? How will STEM communities function if we do, in fact, get the numerical diversity we seek?

## Considering Equity in STEM as a Global Challenge

Comparative and international studies will lead to insights about how access to STEM is stratified by disciplines across multiple contexts. If disciplines are "academic tribes" with their own cultures (Becher, 1989), it may be fruitful to examine STEM communities across countries to better understand what is possible through STEM disciplines even when the national policy contexts create unique circumstances. For example, scholars in the United States may point to the adverse effects of the ACT and SAT standardized tests in preventing access to STEM. Yet, countries like England have similar equity challenges, even though they do not use American standardized tests (Ro, Fernandez, & Alcott, 2018).

Scholars often attribute the lack of equity in STEM to broader, national-level equity challenges. Yet puzzles remain about which countries make the greatest progress in improving equity and inclusion in STEM. Stoet and Geary (2018, 2019) identify what they call a "gender-equality paradox" in STEM; they conclude that the share of women who earn university degrees in STEM fields is lower than expected in countries with higher gender equality. Conversely, countries that have higher gender inequality tend to have more women who graduate from STEM programs. Other scholars challenge Stoet and Geary's analysis, but even they find a correlation between women's attainment in STEM and gender inequality across

75 countries (Richardson et al., 2020). Scholars find a similar pattern with enrollment and completion of massive open online courses (MOOCs) in STEM fields; there are smaller gender gaps in countries with more gender inequality (Jiang, Schenke, Eccles, Xu, & Warschauer, 2018). Certainly, the answer to improving diversity in STEM is not to increase gender inequality. Still, the gender-equality paradox challenges us to consider whether advancing women's rights or improving gender norms and social constructions will in and of itself promote greater participation in STEM.

International and comparative studies can help us test assumptions and conceptual or theoretical frameworks in varying contexts. To make theoretical advances, we must first identify the conditions within which a theory predicts outcomes and then stretch a theory to its limits to understand the contexts where it is no longer useful for understanding a problem (DiMaggio, 1995; Sutton & Staw, 1995). Here is one example of how a comparative study supports interesting findings. Smith and Fernandez (2017) analyze data from Canada and the United States to examine how education helps immigrants gain skills and do well in the labor market. The cross-national study helps the authors identify common relationships between education and skills for immigrants while understanding that outcomes persist based on the conditions of each country's immigration system and labor market.

On a practical level, different countries invest in different types of data collection—and different types of secondary datasets open alternate possibilities for studying research problems. Ro and colleagues (2018) leverage a longitudinal dataset from England to examine how human capital and social capital frameworks are useful to examine the pathway to studying STEM. The study of university students in England confirms that women have lower odds of studying STEM, and it reveals that women have lower odds of studying STEM at the most prestigious universities in England (Ro et al., 2018). Looking globally may help us act locally.

## Adopting Critical Quantitative Methods

Traditionally, researchers who adopt critical race and intersectionality approaches tend to rely on qualitative research methods (Schudde, 2018). In fact, some scholars caution against using quantitative research to examine the intersection of social identities (e.g., Teranishi, 2007). However, higher-education researchers offer recommendations for conducting quantitative research in ways that honor the traditions of critical research. For example, scholars should be reflective and use critical quantitative research to ask thoughtful questions and "unmask inequities" (Rios-Aguilar, 2014, p. 98) related to diversity and equity in STEM.

Scholars increasingly provide examples about how to specify statistical models to support critical quantitative research. For example, researchers use two-way interaction terms to examine the intersection of race and gender to examine student learning outcomes among women in engineering (e.g.,

Ro & Loya, 2015). Others show that effect coding is useful to avoid setting White students as the reference group for comparing minority students' outcomes (Mayhew & Simonoff, 2015). Additionally, Schudde (2018) recommends calculating and plotting marginal effects as one approach for examining intersecting identities.

For brevity's sake, I avoid getting into the statistical details of how to conduct critical quantitative research, but it as integral to continuing to build the scholarly base for STEM education. Echoing Stage and Wells (2014), I "encourage future research using a critical quantitative perspective that will help to shrink the gap between equity-minded research and policy" (p. 3). Anecdotally, I know that too many administrators and policymakers are skeptical of even the best qualitative research. In the future, critical quantitative evidence should complement rich counter-narratives, such as those contained in this volume.

## Supplanting "Resiliency" and "Resilient"

*Resilient* and *resiliency* can suggest that resilience is an attribute or trait—students and organizations have it, or they don't. Scholars tend to challenge deficit views of minoritized students in STEM by characterizing high-achieving students as inherently resilient (high-achieving Latinas are *resilient*; they have *resiliency*). Depictions of resilience often draw on the notion of "resiliency" that is embedded in the idea of community cultural wealth (Yosso, 2005). Looking for resilience among racial minorities and women was a crucial step in flipping the paradigm for scholarly and policy discourse from a deficit-based to an asset-based perspective. But if we think of resilience as a trait, then what do we make of the many women who initially enroll in physical sciences but later major in life or health sciences (George-Jackson, 2011)? If we think of resilience as a trait, are students not resilient when they do not complete their initial majors? Or, is switching majors from one STEM field to another one form of resilience? What about if a low-income, first-generation Latina was studying engineering, dropped out of college, and ultimately graduated with a degree in business—isn't she still showing some sort of resilience? What are different ways to operationalize the concept of resilience, measure it, and positively affect it?

Resilience describes "the achievement of successful adaptation despite developmental risk and adversity" (Gordon & Wang, 1994, p. 191). In that sense, the same student may exhibit resilience when faced with one set of circumstances but not a different set of challenges. We cannot just ascribe desirable outcomes to *resiliency*. When we consider the concept of resilience as a process or as contextually bound, it opens new avenues and opportunities for research. We need to think about the specific qualities or sources of support that are available in a given context and whether they enable adaptation to a particular set of risks or exigencies.

Resilience as a construct is applied to individuals, organizations, and communities (Fernandez & Burnett, 2020; Kamimura, 2020; Sutcliffe & Vogus, 2003; Wang & Gordon, 1994). For all three groups, we should consider how fostering resilience in one context can be a learning opportunity that supports resilience during future challenges (Sutcliffe & Vogus, 2003). In addition to examining resilience among individuals, we should consider how resilience may exist at the department, college, or university level. Achieving greater equity and diversity in STEM will require resilience by multiple actors and organizations.

## Conclusion

This volume is the beginning—and not the culmination—of a research agenda for advancing equity in STEM higher education. As an editor, I had the opportunity and privilege of reviewing each chapter multiple times. The process of compiling this volume allowed me to develop concluding recommendations for continuing the conversation on how to support Latinas—and others—in STEM.

There are connections within this set of recommendations. For instance, we see critical quantitative research methods are crucial for working toward better understanding the double bind. Additionally, students' conceptions of the double bind could change as they reconsider their own gender and racial identities through different life stages or encounter new experiences. Finally, future work should incorporate multiple groups to help examine resilience in STEM as a phenomenon and not a characteristic or trait of any one particular group of underrepresented students. Just as the process of writing six potential directions for future research led me to consider new, cross-cutting opportunities for research, I hope that these comments will inspire ideas for future scholarship among readers.

## References

Becher, T. (1989). *Academic tribes and territories: Intellectual enquiry and the cultures of disciplines*. Milton Keynes, United Kingdom: Society for Research into Higher Education and Open University Press.

Crenshaw, K. W. (1989). Demarginalizing the intersection of race and sex: A Black feminist critique of antidiscrimination doctrine, feminist theory and antiracist politics. In A. Phillips (Ed.), *Feminism and politics* (pp. 314–343). New York, NY: Oxford University Press.

Cross Jr, W. E. (1991). *Shades of Black: Diversity in African-American identity*. Philadelphia, PA: Temple University Press.

DiMaggio, P. J. (1995). Comments on "What theory is not". *Administrative Science Quarterly, 40*(3), 391–397.

Fernandez, F., & Burnett, C. A. (2020). Considering the need for organizational resilience at Hispanic Serving Institutions: A study of how administrators navigate institutional accreditation in Southern states. *International Journal of Qualitative Studies in Education,* 1–17.

Fuller-Rowell, T. E., Burrow, A. L., & Ong, A. D. (2011). Changes in racial identity among African American college students following the election of Barack Obama. *Developmental Psychology, 47*(6), 1608–1618.

Garcia, A., Rincón, B., & Hinojosa, J. K. (2020). "There was something missing": How Latinas construct compartmentalized identities in STEM. In E. M. Gonzalez, F. Fernandez, & M. Wilson (Eds.), *Latina women studying and researching in STEM: An asset-based approach to increasing resilience and retention.* New York, NY: Routledge.

George-Jackson, C. (2011). STEM switching: Examining departures of undergraduate women in STEM fields. *Journal of Women and Minorities in Science and Engineering, 17*(2), 149–171.

Gonsalves, A. J. (2011). Gender and doctoral physics education: Are we asking the right questions? In L. McAlpine & C. Amundsen (Eds.), *Doctoral education: Research-based strategies for doctoral students, supervisors, and administrators* (pp. 117–132). Dordrecht, Netherlands: Springer.

Gonzalez, E., Molina, M., & Turner, S. (2020). Empowering Latina STEM majors at a Tier One public research university and Hispanic-serving institution in Texas: Strategies for success. In E. Gonzalez, F. Fernandez, & M. Wilson (Eds.), *Latina women studying and researching in STEM: An asset-based approach to increasing resilience and retention.* New York, NY: Routledge.

Gordon, E. W., & Wang, M. C. (1994). Epilogue: Educational resilience—challenges and prospects. In M. C. Wang & E. W. Gordon (Eds.), *Educational resilience in inner-city America: Challenges and prospects* (pp. 191–194). Hillsdale, NJ: Lawrence Erlbaum Associates.

Grimes, M. D., & Morris, J. M. (1997). *Caught in the middle: Contradictions in the lives of sociologists from working-class backgrounds.* Westport, CT: Praeger Publishers.

Jiang, S., Schenke, K., Eccles, J. S., Xu, D., & Warschauer, M. (2018). Cross-national comparison of gender differences in the enrollment in and completion of science, technology, engineering, and mathematics massive open online courses. *PLoS One, 13*(9), 1–15.

Kallio, R. E. (1995). Factors influencing the college choice decisions of graduate students. *Research in Higher Education, 36*(1), 109–124.

Kamimura, A. (2020). "Cuida tu casa y deja la ajena": Focusing on retention as a self-perpetuating engine for recruiting Latina faculty in STEM. In E. Gonzalez, F. Fernandez, & M. Wilson (Eds.), *Latina women studying and researching in STEM: An asset-based approach to increasing resilience and retention.* New York, NY: Routledge.

Kim, S., Beverly, S. P., & Ro, H. K. (2020). How many Latinas in STEM benefit from high-impact practices? Examining participation by social class and immigrant status In E. Gonzalez, F. Fernandez, & M. Wilson (Eds.), *Latina women studying and researching in STEM: An asset-based approach to increasing resilience and retention.* New York, NY: Routledge.

Malcom, S., Hall, P., & Brown, J. (1976). *The double bind: The price of being a minority woman in science.* Washington, DC: American Association for the Advancement of Science.

Mayhew, M. J., & Simonoff, J. S. (2015). Non-White no more: Effect coding as an alternative to dummy coding with implications for higher education researchers. *Journal of College Student Development, 56*(2), 170–175.

McMahon, M. (1995). *Engendering motherhood: Identity and self-transformation in women's lives.* New York, NY: Guilford Press.

Mein, E., Muciño Guerra, H., & Herrera-Rocha, L. (2020). Latina undergraduates in engineering/computer science on the US-Mexico border: Identity, social capital, and persistence. In E. Gonzalez, F. Fernandez, & M. Wilson (Eds.), *Latina women studying and researching in STEM: An asset-based approach to increasing resilience and retention.* New York, NY: Routledge.

National Science Foundation (2017). *Women, minorities, and persons with disabilities in science and engineering, 2015.* Washington, DC: National Center for Science and Engineering Statistics, National Science Foundation.

Ong, M., Wright, C., Espinosa, L. L., & Orfield, G. (2011). Inside the double bind: A synthesis of empirical research on undergraduate and graduate women of color in science, technology, engineering and mathematics. *Harvard Education Review, 81*(2), 172–201. https://doi.org/10.17763/haer.81.2.t022245n7x4752v2

Renn, K. A. (2000). Patterns of situational identity among biracial and multiracial college students. *The Review of Higher Education, 23*(4), 399–420.

Richardson, S. S., Reiches, M. W., Bruch, J., Boulicault, M., Noll, N. E., & Shattuck-Heidorn, H. (2020). Is there a gender-equality paradox in science, technology, engineering, and math (STEM)? Commentary on the study by Stoet and Geary (2018). *Psychological Science, 31*(3), 338–341.

Rios-Aguilar, C. (2014). The changing context of critical quantitative inquiry. *New Directions for Institutional Research, 2013*(158), 95–107.

Ro, H. K., Fernandez, F., & Alcott, B. (2018). Social class, human capital, and enrollment in STEM subjects at prestigious universities: The case of England. *Educational Policy.* doi:0895904818813305

Ro, H. K., & Loya, K. I. (2015). The effect of gender and race intersectionality on student learning outcomes in engineering. *The Review of Higher Education, 38*(3), 359–396.

Schudde, L. (2018). Heterogeneous effects in education: The promise and challenge of incorporating intersectionality into quantitative methodological approaches. *Review of Research in Education, 42*(1), 72–92.

Smith, W. C., & Fernandez, F. (2017). Education, skills, and wage gaps in Canada and the United States. *International Migration, 55*(3), 57–73.

Snyder, T. D., & Dillow, S. A. (2012). *Digest of education statistics, 2012.* National Center for Education Statistics. Retrieved from https://nces.ed.gov/programs/digest/d12/tables/dt12_310.asp

Stage, F. K., & Wells, R. S. (2014). Critical quantitative inquiry in context. *New Directions for Institutional Research, 2013*(158), 1–7.

Stoet, G., & Geary, D. C. (2018). The gender-equality paradox in science, technology, engineering, and mathematics education. *Psychological Science, 29*(4), 581–593.

Stoet, G., & Geary, D. C. (2019). Corrigendum: The gender-equality paradox in science, technology, engineering, and mathematics education. *Psychological Science, 31*(1), 110–111.

Strayhorn, T. L., Long III, L., Kitchen, J. A., Williams, M. S., & Stenz, M. E. (2013). Academic and social barriers to Black and Latino male collegians' success in engineering and related STEM fields (Paper ID #8199). Proceedings from 2013 ASEE Annual Conference and Exposition, Atlanta, GA.

Sutcliffe, K. M., Vogus, T. J. (2003). Organizing for resilience. In K. S. Cameron, J. E. Dutton, & R. E. Quinn (Eds.), *Positive organizational scholarship: Foundations of a new discipline* (pp. 94–110). San Francisco: Berrett-Koehler.

Sutton, R. I., & Staw, B. M. (1995). What theory is not. *Administrative Science Quarterly*, *40*(3), 371–384.

Teranishi, R. T. (2007). Race, ethnicity, and higher education policy: The use of critical quantitative research. *New Directions for Institutional Research*, *2007*(133), 37–49.

Wang, M. C., & Gordon, E. W. (Eds.). (1994). *Educational resilience in inner-city America: Challenges and prospects*. Hillsdale, NJ: Lawrence Erlbaum Associates.

Yosso, T. J. (2005). Whose culture has capital? A critical race theory discussion of community cultural wealth. *Race Ethnicity and Education*, *8*(1), 69–91.

# Index

Note: *Italicized* page numbers indicate a figure on the corresponding page. Page numbers in **bold** indicate a table on the corresponding page.

gender-equality paradox 196–197
gender equity lawsuits 62
gender identity 103, 187–188, 194
gender-neutral restrooms 65
gender stereotypes 26, 117–118, 125
geographical identity 33
global-focused academic experiences 80, 81, 85, *85*
Google Scholar 13, 14
Grace Hopper Celebration 36

health sciences 185
high-impact practice (HIP): analytical sample of 80; Capstone projects *81*, 81–82; data on 79; effects of 78–79; findings 80–86, *81*; introduction to 6, 75–77; participation in 100; recommendations for future research 87–88; social and resistant cultural capital 77–78; summary of 86–87; variables with 79–80
Hispanic-serving institution (HSI) 78, 101, 132, 165, 168–169; *see also* Research 1 (R1) Doctoral Hispanic-serving institution
honors programs 80, *82*, 82–84
Howard Hughes Medical Institute 49
Huber, Pérez 120

identity: collegiality and identity intersectionality 59; cultural identity 33, 98–103, 186; development of 2; disciplinary role identity 147; ethnic identity 19–20, 26, 104; funds of identity 27, 32–37; gender identity 103, 187–188, 194; intersectional identities 123–124; multidimensional identities 182, 188–190; oppression and 148–149; self-identity 103–107, 111; sense-making of 29; SES and formation of 104; social identity 12, 33, 51, 62, 67, 77, 83–84, 182–189, 197; sociocultural perspective on 132–133; student identities 119–120, 158; *see also* agentic identity of engineering student; compartmentalized identities in STEM; computing identity development
identity-based student organizations 19–20
immigrant generation status 83, 84–85

individual level computing identity development 27–32, *28*
information-sharing pipelines 87
institutional identity 33
internalized funds of knowledge 33
internal resilience factors 98
internships 80, *84*, 84–85
interpersonal relationships with STEM communities 116
intersectional identities: in engineering climate 123–124; equity and 196–197; gender and race 194–195; introduction to 193; qualitative research methods 197–198; resiliency *vs.* resilient 198–199; summary of 199
intersectionality, defined 34
intrinsic motivation 166

Joint Working Group on Improving Underrepresented Minorities Persistence in Science, Technology, Engineering, and Mathematics 49
JSTOR 13, 14

knowledge accumulation 65

language acquisition 79
Latina ethnicity and leadership capacity 166–167
Latina faculty retention strategies: building to critical mass 65–66; "children in the hallways" strategy 67–68; "chilly climate" of faculty of color 60; "*cuida tu casa y deja la ajena*" 57, 71–72; current state of 57–61; enhancing resilience 65–68, 71; *familia* concept 61, 68–71; gatekeepers in 58–61; gender-neutral restrooms 65; healthy environment of 60–61; inclusion promotion 63–65; introduction to 57; revolving door syndrome 60; role-modeling strategy 69–70; self-perpetuating strategies 61–71; "Supreme Court rules" strategy 67; tenure-track faculty 68, 70–71; transparency practices 59, 63–65; university department concepts 62–63
Latina resilience 1–7, *3*, 109–110
Latina/o resilience model 3–4
Latinx Cultural Center 185
leadership capacity of Latina STEM students: *Coming Out of the Shadows* theme 171–173; discussion on 175–177; *Embracing Their Identity*

For Product Safety Concerns and Information please contact our EU
representative GPSR@taylorandfrancis.com
Taylor & Francis Verlag GmbH, Kaufingerstraße 24, 80331 München, Germany